THE ELECTRIC PICKLE

THE ELECTRIC PICKLE

50 Experiments from the PERIODIC TABLE, from Aluminum to Zinc

JOEY GREEN

WARNING: A responsible adult should supervise any young reader who conducts the experiments in this book to avoid potential dangers and injuries. The author has conducted every experiment in this book and has made every reasonable effort to ensure that the experiments are safe when conducted as instructed. However, the author and publisher of this book disclaim all liability incurred in connection with the use of the information contained in this book.

Published by Chicago Review Press Incorporated
814 North Franklin Street
Chicago, Illinois 60610
ISBN 978-1-61373-959-4

Library of Congress Cataloging-in-Publication Data
Names: Green, Joey, author.
Title: The electric pickle : 50 experiments from the periodic table, from aluminum to zinc / Joey Green.
Description: Chicago, Illinois : Chicago Review Press, [2018] | Includes bibliographical references.
Identifiers: LCCN 2017001557 (print) | LCCN 2017001881 (ebook) | ISBN 9781613739600 (pdf) | ISBN 9781613739624 (epub) | ISBN 9781613739617 (Kindle) | ISBN 9781613739594 (pbk. : alk. paper)
Subjects: LCSH: Chemistry—Experiments. | Chemical elements. | Electricity—Experiments. | Periodic table of the elements.
Classification: LCC Q164 (ebook) | LCC QD38 .G74 2018 (print) | DDC 540.78—dc23
LC record available at https://lccn.loc.gov/2017001557

Cover design: Andrew Brozyna
Cover and interior images: Debbie Green
Interior design: Jonathan Hahn

Printed in the United States of America
5 4 3 2 1

"If it squirms, it's biology.
If it stinks, it's chemistry.
If it doesn't work, it's physics.
And if you can't understand it, it's mathematics."

—Magnus Pyke, British scientist

CONTENTS

Introduction

In the mid-19th century, airship inventors and builders filled the first airships—more commonly known as dirigibles, zeppelins, and blimps—with hydrogen, the lightest gas known, causing the airships to rise high into the sky. Weighing only about 7 percent of an equal volume of air, hydrogen can be produced by the electrolysis of water, in which an electric current breaks down the water into its two elements: hydrogen and oxygen. Hydrogen, however, is extremely flammable. In 1937, while approaching Lakehurst, New Jersey, on a flight from Germany, the Hindenburg, the largest airship in its day, exploded and burst into flames in the first air disaster recorded on film. The crash destroyed the Hindenburg in 34 seconds, ended the use of airships for regular passenger service, and raised an obvious question: Why not use non-explosive helium instead?

In 1896, French physicist Henri Becquerel discovered that minerals containing uranium emitted X-rays. Physicist Marie Curie, having received her master's degree from the Sorbonne, decided to investigate the uranium rays, setting up a laboratory in a storeroom in the Paris Municipal School, where her husband, Pierre Curie, was a professor. After discovering that compounds containing the uncommon element thorium also emanated rays, Marie invented the term *radioactivity* to describe the atomic property of the two chemical elements. Together with her husband, Marie discovered two more radioactive elements: polonium and radium. In 1903 she became the first woman to receive a doctorate in France and the first woman to receive the Nobel Prize. Refusing to believe that radiation was harmful, Marie died of leukemia on July 4, 1934, plausibly due to exposure to high levels of radiation emitted by the radioactive elements she had been studying during the previous four decades.

Now I'm not advocating that you play with radioactive elements or set fire to a balloon filled with hydrogen, but I can tell you that the most dangerous science experiments—launching rockets, mixing up weird chemicals, splitting atoms, or playing with a particle accelerator—do tend to be the most exciting. Until you've walked on eggs, powered a rocket with Diet Coke and Mentos, or turned an ordinary pickle into a lightbulb that throws off hot orange sparks, you really haven't lived life to the fullest. There's something strangely exhilarating about shrinking a potato chip bag to the size of a domino, microwaving a lightbulb, or creating an explosive fireball with cornstarch.

Is there any way to justify this downright kooky behavior? Oddly, yes. Scores of amazingly intelligent people who have pursued careers in science share a love for blowing up things and putting themselves in harm's way. In the 1920s, when rocket scientist Robert H. Goddard saw his first four attempts at launching a rocket blow up on the launch pad, I'm guessing his first thought was something along the lines of "Cool!" In 1899, Serbian American inventor Nikola Tesla sent a heavy electrical current through the coils wrapped around a tower 80 feet in height and up a 142-foot-tall metal pole topped by a 3-foot-diameter copper ball—producing bolts of artificial lightning that measured up to 135 feet long, generating thunder heard fifteen miles away, and inadvertently knocking out the electrical power station in Colorado Springs. In 1924, at the age of 12, future aerospace engineer Wernher von Braun strapped six fireworks rockets to his toy wagon, ignited them, and shot off across a crowded Berlin street—where a police officer apprehended him.

The eccentric experiments in this book spark the imagination, bring science to life, create excitement about how the world works, demonstrate the undeniable beauty and importance of science, and enhance your understanding of the most common elements in the periodic table—the unique substances that cannot be broken down or made into anything simpler by chemical reactions. Scientists have identified more than 110 elements, and the experiments in this book involve only those elements (or a compound that includes the specific element) that are easily available and safe to handle. In some cases the experiments focus on the basic attributes of a particular element rather than directly on the element itself. Unfortunately, conducting these experiments also comes with the danger that society might label you a geek, weirdo, or nutjob. The good news? You'll be in excellent company. Alexander Graham Bell, Thomas Alva Edison, Robert Goddard, Albert Einstein—at one time they were all considered ditzy crackpots.

Of course, everything in life is potentially dangerous. Simply flipping through the pages of this book can give you a painful paper cut, and boiling a pot of water can cause third-degree burns, should you accidentally bump the pot, spilling boiling water all over yourself. Remember, accidents happen, and any science experiment, regardless of its simplicity or how precisely you follow the instructions, can be unpredictable. Even if you're working with harmless ingredients, take the proper precautions by wearing appropriate safety equipment, following the delineated procedures, securing adult supervision when necessary, and concentrating on what you are doing. While I have personally conducted all the experiments in this book and thoroughly reviewed the described instructions, please be cautious when performing the experiments. I'm sure Marie Curie would have loved conducting every project with reckless abandon. But I urge you to wear gloves to protect yourself from possible paper cuts.

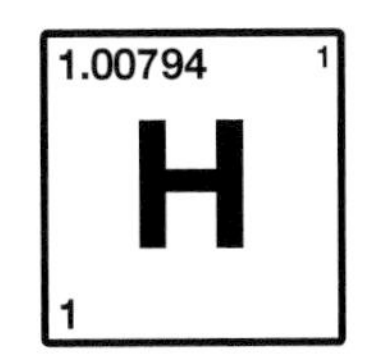

HYDROGEN

1. Exploding Hydrogen Bubbles

SAFETY FIRST

This experiment should be performed outside.

- Rubber gloves
- Safety goggles
- Respirator mask
- Earplugs

WHAT YOU NEED

- Plastic shoebox container
- Water
- Measuring spoons
- Dishwashing liquid
- Funnel
- Measuring cup
- Wine bottle, clean and empty
- Scissors
- Aluminum foil (12 by 12 inches)
- Ruler
- Crystal Drano (containing sodium hydroxide)
- Balloon
- Butane barbecue lighter

WHAT TO DO

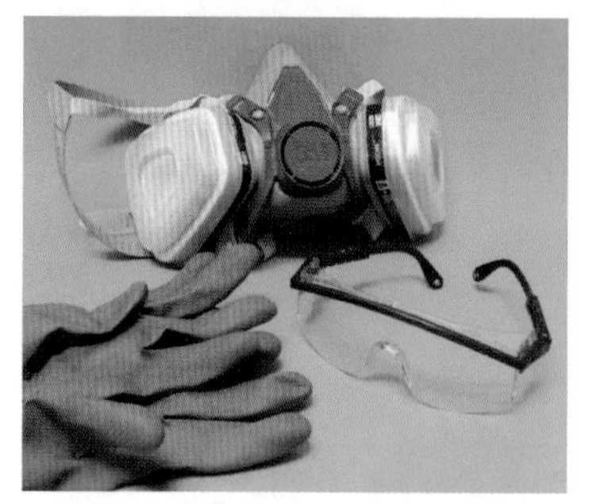

1. Fill the plastic shoebox container halfway with water. Add 1 teaspoon dishwashing liquid and mix well, without creating many bubbles. Set aside for later.
2. Working outdoors, use the funnel to pour 2 cups of water into the wine bottle.
3. Make 20 small balls from the aluminum foil (about ½ inch in diameter) and set aside for the moment.
4. Wearing rubber gloves, safety goggles, and a respirator mask, carefully add 3 tablespoons Crystal Drano into the bottle. Carefully swirl the bottle to dissolve the Drano.
5. Drop the 20 small balls of aluminum foil into the bottle.
6. Stretch the balloon well and immediately place the neck of the balloon over the mouth of the bottle. Wait approximately 10 minutes to allow the balloon to inflate as large as it can get. (If the solution doesn't give off enough gas to fill the balloon, add more aluminum; if the glass gets too hot, you've used too much aluminum.)

7. When the balloon is full, place the earplugs in your ears.
8. Squeezing the neck of the balloon to prevent any gas from escaping from the balloon, remove the balloon from the mouth of the bottle. **Do not breathe the vapors from the bottle.**
9. Submerge the mouth of the balloon in the soapy solution in the plastic container, and carefully allow the gas in the balloon to empty into the solution, creating soap bubbles.
10. Using the butane barbecue lighter, light the flame and touch it to the soap bubbles.
11. Carefully dispose of the remaining solution in the bottle.

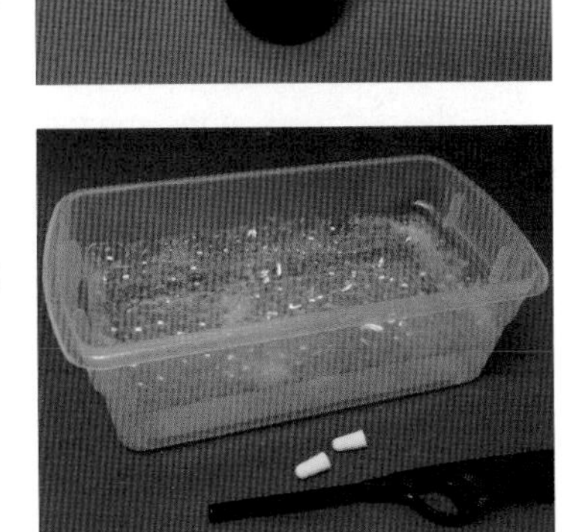

WHAT HAPPENS

The soap bubbles explode.

WHY IT WORKS

The sodium hydroxide in the lye (Drano) reacts with the aluminum to produce aluminum chloride and hydrogen. The chemical reaction produces heat, warming the glass bottle. Releasing the hydrogen into the soapy water creates soap bubbles filled with explosive hydrogen gas.

WACKY FACTS

- The *Hindenburg*, one of the largest airships ever built, burst into flames in 1937 over Lakehurst, New Jersey, when its hydrogen-filled bag exploded, killing 36 people. Today airships are filled with helium.
- Hydrogen is used as rocket fuel because the combustion reaction between hydrogen and oxygen propels the exhaust gas (primarily water vapor) out of the rocket's engine at 7,910 miles per hour—creating enormous thrust that can lift the 4.4-million-pound space shuttle into orbit.

- The first hydrogen bomb, detonated in Enewetak Atoll in the Pacific in 1952, used fission to cause the fusion of the nuclei of two hydrogen atoms—yielding an explosion equivalent to 10 million tons of TNT.
- If a hydrogen atom were the size of a golf ball, a golf ball would be the size of the Earth.

Without Hydrogen, We'd Be Drinking Oxygen

- Hydrogen is one of two elements found in water (H_2O). Each molecule of water consists of two hydrogen atoms bonded to one oxygen atom.
- Hydrogen is the most abundant element in the universe, making up more than 90 percent of all the atoms.
- Hydrogen was named after the Greek words *hydro*, meaning "water," and *genes*, meaning "creator."
- Scientists believe that hydrogen was one of three elements produced in the Big Bang. The other two are helium and lithium.
- Hydrogen reacts explosively with the elements oxygen, chlorine, and fluorine.

4.002602 2
He
2

2. Hexed Helium Balloon

WHAT YOU NEED

- Helium balloon with string
- Car, van, or enclosed truck

WHAT TO DO

1. Place the helium balloon in the backseat of your car, van, or truck. If possible, remove the backseats from the van and tie the end of the balloon's string to a bolt on the floor.
2. Drive to an empty road or parking lot.
3. Accelerate quickly (without breaking the speed limit).
4. As you feel the force pushing you backward during acceleration, observe the motion of the balloon.
5. Apply the brakes, and noting the force you feel pushing you forward, observe the motion of the balloon.

WHAT HAPPENS

When you accelerate the vehicle, the helium balloon moves forward in the direction of acceleration. When you apply the brakes, the helium balloon moves backward.

WHY IT WORKS

The helium, weighing less than air, causes the balloon to float upward.

An object at rest stays at rest unless acted upon by another force. You are the object at rest inside the vehicle. When you accelerate the vehicle, you feel the force of the seat pushing you forward and accelerating along with the car. Similarly, when you accelerate, the air inside the vehicle wants to remain stationery, but without a seat to hold them in place, the molecules of air move toward the back of the car, creating more air pressure in the back of the car than the front of the car. This density gradient in the air causes the helium, being less dense than air, to move forward in the car, where the air molecules are less dense.

When you hit the brake, the air inside the vehicle continues moving forward, creating more air pressure in the front of the vehicle than in the back. This reversed density gradient causes the helium to move backward toward the less dense air.

WACKY FACTS

- A helium balloon floats in the air only if the weight of the helium combined with the weight of the balloon is lighter than the weight of the air it displaces.
- Helium weighs 0.1785 grams per liter. Air weighs approximately 1.25 grams per liter.
- On July 5, 2015, Daniel Boria of Calgary, Alberta, Canada, attached 110 helium balloons (each with diameter of six feet) to a lawn chair, rose into the sky above the clouds, and parachuted into an empty field. Police arrested Boria and charged him with one count of mischief causing danger to life, referring to the lawn chair, which police said could cause damage or hurt someone when the balloons pop and the chair falls to the ground.

Helium on the Rise

- French astronomer Pierre Janssen and English astronomer Sir Norman Lockyer discovered helium using spectral analysis of sunlight after a solar eclipse in 1868.
- Being lighter than air, helium is commonly used to fill airships, balloons, and blimps.
- Helium does not burn or react with other chemicals, making it relatively safe.
- Inhaling helium temporarily raises the pitch of a person's voice. Helium is nontoxic. However, inhaling too much helium can cause death by asphyxiation due to oxygen deprivation.
- The United States produces most of the world's helium (approximately 75 percent), followed by Qatar (roughly 14 percent).
- Helium is the second most abundant element in the universe.
- In 1895, Scottish chemist Sir William Ramsay and Swedish chemists Nils Langlet and Per Theodor Cleve first found helium on Earth in the mineral clevite.
- English astronomer Sir Norman Lockyer named helium after the Greek god of the sun, Helios.
- The only element lighter than helium is hydrogen.
- Helium does not combine easily with other elements, and no known compounds contain helium.

LITHIUM

6.941	2 1
Li	
3	

3. Horrible Hot Dog

SAFETY FIRST

- Do not leave this experiment unattended.
- Keep this experiment away from animals.

WHAT YOU NEED

- Hot dog
- Ceramic plate (*not* paper or plastic)
- Sharp kitchen knife
- Lithium battery, 3 volt

WHAT TO DO

1. Place the hot dog on a plate, and carefully use the kitchen knife to slice a small slot in the center of the hot dog.
2. Insert the lithium battery in the slot.
3. Observe the hot dog after 1 hour.
4. After 12 hours, slice the hot dog in half on a diagonal at the spot where you inserted the lithium battery. Remove the battery and observe.

WHAT HAPPENS

In less than a minute, the hot dog begins sizzling and bubbling. Within an hour the battery burns the edges of the slot and the hot dog continues sizzling. After 2 hours, when you slice the hot dog in half and remove the battery, you'll see a hole starting to burn through the hot dog.

WHY IT WORKS

The fluids in the hot dog allow the electric current from the battery to pass to the tissues of the meat. The electrical current generates heat and kills the cells in the tissue. Similarly, if a child or pet swallows one of these batteries, the electrical current, flowing through saliva or other bodily fluids, can severely damage or perforate the pharynx, esophagus, stomach, or small intestine—within 15 to 30 minutes.

WACKY FACTS

- Electronic greeting cards, key fobs, wristwatches, remote controls, and garage door openers frequently contain tiny, shiny, button-shaped lithium batteries, which small children and pets can easily swallow.
- A disc-shaped battery inserted into a nostril or ear canal sends an electric current between the positive and negative poles of the battery, essentially electrocuting the lining inside the nose or ear canal, leading to tissue damage and ultimately necrosis (tissue death).
- Warnings on battery packaging state that swallowing a lithium battery can lead to serious injury or death in less than two hours due to chemical burns and potential perforation of the esophagus. However, this experiment illustrates that the battery can seriously damage the hot dog in roughly 20 minutes—far less than 2 hours.
- Keep battery-powered devices and spare batteries out of the reach of children and pets to prevent them from accidentally swallowing the batteries or inserting a battery into their nose or ear.

Getting Pithier with Lithium

- Lithium is the lightest metal, with a density about half that of water. If lithium did not react intensely with water (which it does, forming lithium hydroxide and highly flammable hydrogen), it would float.
- Lithium is a metal soft enough to cut with a knife.
- Swedish chemist Johan Arfvedson discovered lithium in 1817. The following year, Swedish chemist William Brande and English chemist Sir Humphry Davy, working independently, isolated the element.
- The name *lithium* is derived from Greek word *lithos*, meaning "stone."
- Lithium occurs in most igneous rocks, but it does not occur free in nature.
- The lithium used as a psychiatric medication is not pure lithium but rather a compound, most commonly lithium carbonate, lithium citrate, and lithium orotate.

BORON

10.811 2 3
B
5

4. Green Tornado Fire

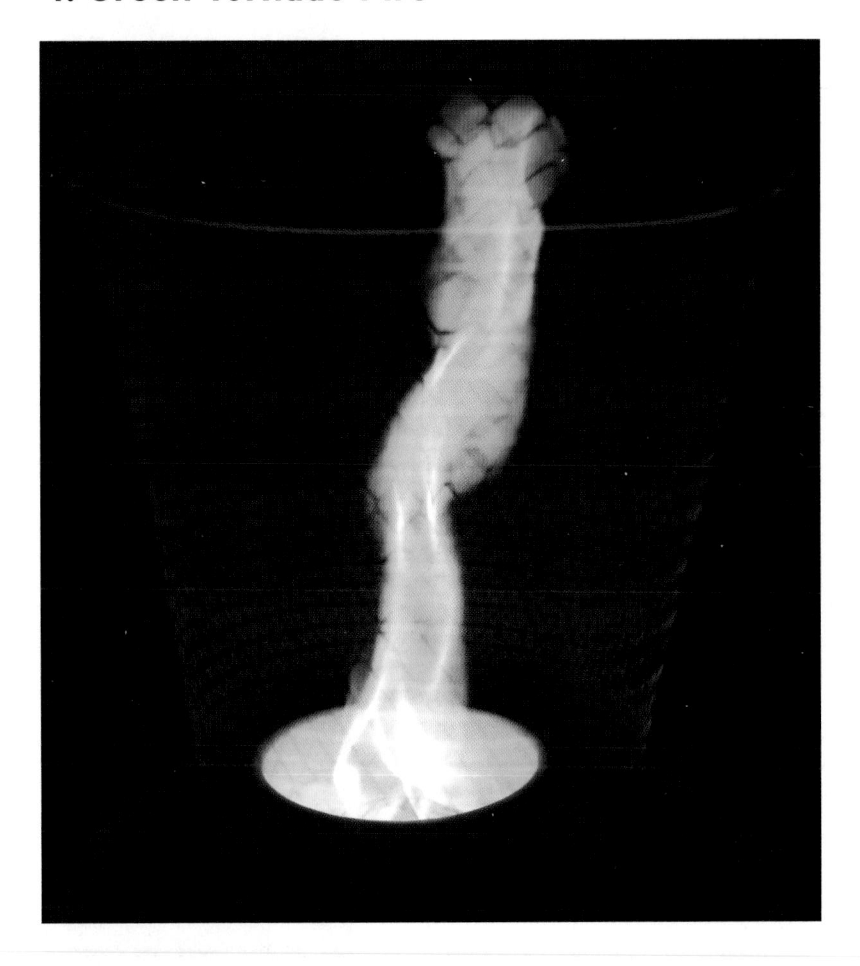

SAFETY FIRST

This experiment should be performed outside at night on a solid surface.

- Safety goggles
- Respirator mask
- Work gloves

WHAT YOU NEED

- Round wire-mesh wastebasket, 8-inch diameter
- Lazy Susan plastic turntable, 9-inch diameter
- Ceramic bowl, 5-inch diameter
- Boric acid (available at hardware stores)
- Measuring spoons
- Heet Gas-Line Antifreeze and Water Remover (available at automotive supply stores)
- Butane barbecue lighter
- Ceramic saucer or plate (optional)

WHAT TO DO

1. Set the wastebasket on the lazy Susan on level pavement outside at night.
2. Practice spinning the contraption to make sure you can do so without causing the wastebasket and lazy Susan to topple over.
3. Place the ceramic bowl inside on the bottom of the wastebasket.
4. Wearing safety goggles, a respirator mask, and work gloves, place a tablespoon of boric acid in the bottom of the ceramic bowl.
5. Without breathing the fumes, carefully pour a few teaspoons of the Heet solution on top of the boric acid in the ceramic bowl. Replace the cap on the Heet and place the bottle a safe distance from the experiment. Stir carefully with the measuring spoon for 15 seconds.

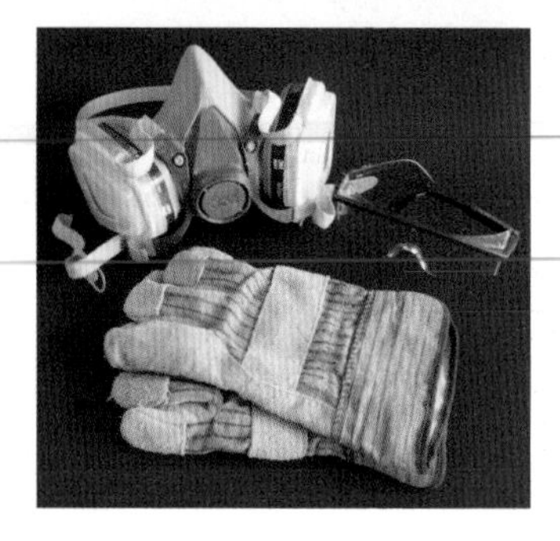

6. Using a butane barbecue lighter, carefully ignite the mixture in the bowl. Be aware that the methanol in the Heet solution is extremely flammable and produces intense heat.
7. Carefully spin the lazy Susan, without splattering any of the mixture.
8. When the flame extinguishes itself, allow the bowl to cool for 5 minutes before using the work gloves to pick up and handle the container.

WHAT HAPPENS

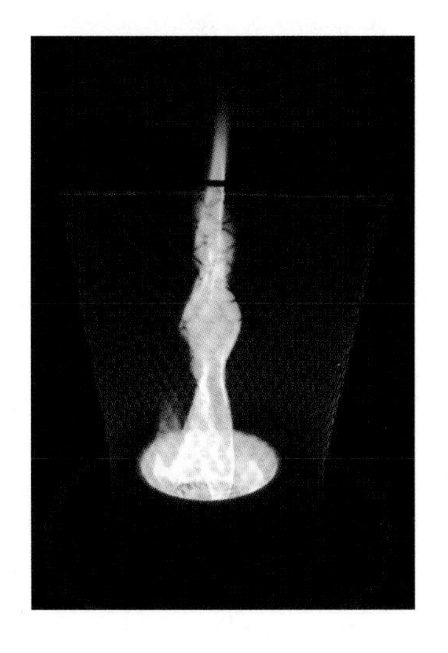

After roughly 10 seconds, green flames rise from the bowl in the shape of a tornado funnel. The flaming alcohol extinguishes itself, or you can put out the flame by covering the bowl with a ceramic saucer or plate.

WHY IT WORKS

When ignited by the methanol in the Heet to the proper temperature, the boron in the boric acid yields green flames. Spinning the lazy Susan forces the air column created by the wire basket into a vortex, sending the green flames upward into a tall cylindrical shape.

WACKY FACTS

- Drinking a single can of diet soda containing 200 milligrams of aspartame, an artificial sweetener made from aspartic acid and the methyl ester form of phenylalanine, produces 20 milligrams of methanol in the body. The gastrointestinal tract hydrolyzes aspartame into its components: methanol, phenylalanine, and aspartic acid. The amount of methanol produced during the digestion of aspartame is small compared to that which is obtained from everyday foods. For example, an eight-ounce glass of tomato juice provides more than three times as much methanol as an eight-ounce glass of a beverage sweetened with aspartame.

- Methanol is naturally found in many fruits, vegetables, and fruit juices in low levels. None provide enough methanol to cause toxicity.
- Fire jugglers and fire spinners frequently use boric acid dissolved in methanol to create a deep-green flame.
- Using isopropyl alcohol in this experiment rather than Heet Gas-Line Antifreeze and Water Remover produces flames that alternate from orange to blue to green.
- Pouring methanol into a hot container may cause spontaneous combustion. Do not store methanol near an open flame or possible ignition source.

Nothing Boring About Boron

- In 1808, French chemists Joseph-Louis Gay-Lussac and Louis-Jacques Thénard first isolated boron by combining boric acid with potassium. That same year English chemist Sir Humphry Davy did the same—independently. Today, scientists typically obtain boron by heating borax with carbon.
- Boron gets its name from the Arabic word *buraq* and the Persian word *burah*, words for the compound borax.
- Boron forms several commercially important compounds, including boric acid (used as a flame retardant and insecticide) and borax (utilized in laundry detergents and as a mild antiseptic).
- The compound boron nitrate is almost as hard as diamond.

5. Rubber Blubber Slime

WHAT YOU NEED

- 4-ounce bottle of Elmer's School Glue Gel
- 2 large glass mixing bowls
- Water
- Green food coloring
- 2 spoons
- Measuring cup
- Measuring teaspoon
- 20 Mule Team Borax
- Ziplock storage bag or airtight container

WHAT TO DO

1. Empty the bottle of Elmer's School Glue Gel into the first bowl.
2. Fill the empty glue bottle with water, and then pour it into the bowl of glue.
3. Add 10 drops of food coloring, and stir well.
4. In the second bowl, mix 1 teaspoon of 20 Mule Team Borax with 1 cup of water. Stir until the powder dissolves.

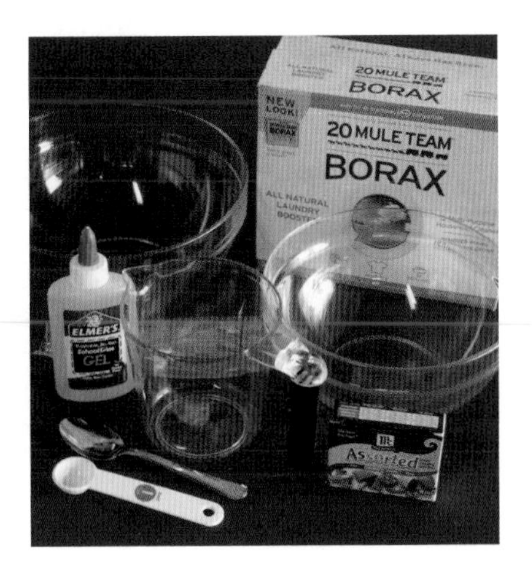

5. Slowly pour the colored glue into the bowl containing the borax solution, stirring as you do so.
6. Remove the thick glob that forms, and knead the glob with your hands until it feels smooth and dry.
7. Discard the excess water remaining in the bowl.
8. Store the Rubber Blubber Slime in the ziplock bag or airtight container.

WHAT HAPPENS

The resulting soft, pliable, rubbery, translucent glob snaps if pulled quickly, stretches if pulled slowly, and slowly oozes to the floor if placed over the edge of a table.

WHY IT WORKS

The polyvinyl acetate molecules in the glue act like invisible bicycle chains drifting around the water. The borax molecules (sodium borate, a compound of sodium and boron) act like little padlocks, locking the chain links together wherever they touch the chain. The locks and chains form an interconnected "fishnet," and the water molecules act like fish trapped in the net.

WACKY FACTS

- Rubber Blubber Slime is a non-Newtonian fluid—a liquid that does not abide by any of Sir Isaac Newton's laws on how liquids behave. Quicksand, gelatin, and ketchup are also non-Newtonian fluids.
- Increasing the amount of borax in the second bowl makes the slime thicker. Decreasing the amount of borax makes the slime slimier and oozier.
- A non-Newtonian fluid's ability to flow can be changed by applying a force. Pushing or pulling on the slime makes it temporarily thicker and less oozy.
- The brand 20 Mule Team Borax is named for the 20-mule teams used during the late 19th century to transport borax 165 miles across the desert from Death Valley to the nearest train depot in Mojave, California. The 20-day round trip started 190 feet below sea level and climbed to an elevation of over 4,000 feet before it was over.
- Between 1883 and 1889 the 20-mule teams hauled more than 20 million pounds of borax out of Death Valley. During this time, not a single animal was lost nor did a single wagon break down.
- Today it would take more than 250 mule teams to transport the borax ore processed in just one day at Pacific Coast Borax Company's modern facility in the Mojave Desert.
- Although the mule teams were replaced by railroad cars in 1889, 20-mule teams continued to make promotional and ceremonial appearances at events ranging from the 1904 St. Louis World's Fair to President Woodrow Wilson's inauguration in 1917. They won first place in the 1917 Pasadena Rose Parade and attended the dedication of the San Francisco Bay Bridge in 1937.
- Borax deposits in Death Valley were abandoned when richer deposits were found elsewhere in the Mojave Desert, turning mining settlements into ghost towns that now help make the region a tourist attraction.
- According to legend, borax was used by Egyptians in mummification.
- In the furniture business, the word *borax* signifies cheap, mass-produced furniture.
- Once proclaimed to be a "miracle mineral," 20 Mule Team Borax was used to aid digestion, keep milk sweet, improve the complexion, remove dandruff, and even cure epilepsy.

12.0107	2 4
C	
6	

6. Bizarre Gummy Bears

WHAT YOU NEED

- Clear drinking glass
- Water
- Gummy bears
- Spoon

WHAT TO DO

1. Fill a clear drinking glass halfway with water.
2. Drop one gummy bear into the glass of water.
3. Observe.
4. Set the glass aside in a safe place where it cannot be disturbed for 12 hours to 3 days.

5. Observe the gummy bear again.
6. Compare a fresh gummy bear to the gummy bear in the water.
7. Using a spoon, remove the gummy bear from the water, and test to see if the gummy bear bounces on the countertop.

WHAT HAPPENS

The gummy bear fills with water and grows larger.

WHY IT WORKS

During the manufacturing process, the warm, liquid solution of sugars (compounds of carbon, hydrogen, and oxygen), gelatin (another compound of carbon, hydrogen, and oxygen), water, and flavoring used to make gummy bears cools and hardens. Water is drawn from the solution, causing the gelatin to form a solid matrix that traps a small amount of water. When a gummy bear sits in a glass of water, the molecules of water in the glass (the less concentrated solution) pass through the semipermeable membrane of the gummy bear to equalize the more concentrated molecules of water in the gummy bear—to balance the concentrations on each side of the membrane. This process is known as osmosis.

WACKY FACTS

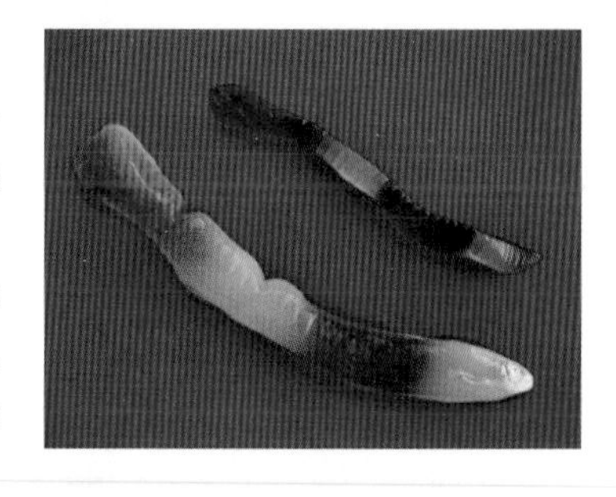

- You can also conduct this experiment with gummy worms, which will double in size.
- An area with a higher concentration of a solution is called *hypertonic*. Areas with lower concentrations are called *hypotonic*.
- A process vital to life, osmosis maintains equilibrium inside and outside a cell. However, a high concentration of salt outside a plant causes the water inside the plant cells to diffuse outside the plant, causing the plant cells to shrink.

7. Flaming Black Snakes

SAFETY FIRST

This experiment should be performed outside on a concrete surface.

- Safety goggles
- Rubber gloves

WHAT YOU NEED

- Confectioner's (powdered) sugar
- Measuring spoon
- Baking soda
- Mixing bowl
- 4 cups of play sand
- Metal pie pan
- Spoon
- Isopropyl alcohol, 91 percent
- Butane barbecue lighter

WHAT TO DO

1. Mix 4 tablespoons of confectioner's sugar with 1 tablespoon of baking soda in a mixing bowl.
2. Pour a mound of sand in the pie pan, and using the spoon, make a small depression (or crater) in the center of the mound.
3. Pour ¼ cup of isopropyl alcohol into the depression.
4. Pour the sugar and baking soda mixture into the depression.
5. Wearing safety goggles and sturdy rubber gloves (and working outside), use the butane barbecue lighter to ignite the sugar and baking soda mixture. Do not add alcohol to the burning snakes. Doing so risks setting the stream of alcohol and the surrounding area—and you—on fire.
6. Stand back and observe.

WHAT HAPPENS

The alcohol ignites, and black bubbles emerge from the sugar and baking soda mixture. Shortly afterward, the black bubbles enlarge and turn into columns of black ash that continuously emerge from the sand and flames, as if from nowhere, looking like black snakes. If you use the spoon to clear away the black snakes from the white mixture while the flames continue, more black snakes will form.

WHY IT WORKS

The flame produced by isopropyl alcohol, reaching a maximal temperature of 680°F, chars the sugar, leaving behind a black residue that is primarily carbon. The heat simultaneously causes the baking soda (sodium bicarbonate) to decompose to form sodium carbonate, water vapor, and carbon dioxide gas. As small bubbles of carbon dioxide gas are released, they form pockets in the charred sugar, which causes it to expand and grow like a snake, producing a twisting tube of black carbon ash.

WACKY FACTS

- The growing snakes smell like burned marshmallows.
- Another way to produce Flaming Black Snakes is by igniting the

chemical compound mercury(II) thiocyanate, a white powder, which quickly produces ash resembling a large, coiling serpent. German chemist Friedrich Wöhler discovered this attribute of mercury(II) thiocyanate in 1821. In 1865, a letter to the *Pharmaceutical Journal* reported that the chemical was being sold as a toy under the name "Pharaoh's Serpent" in Paris. "It consists of a little cone of tinfoil, containing a white powder, about an inch in height, and resembling a pastille," explained the writer. "This cone is to be lighted at its apex, when there immediately begins issuing from it a thick, serpent-like coil, which continues twisting and increasing in length to an almost incredible extent." Several countries banned the product after scientists discovered the substance contains two toxic ingredients—mercury and sulphocyanic acid—and several children died after accidentally ingesting the mercury and inhaling mercury fumes.

- Today, commercially available black snake fireworks that produce an effect similar to mercury(II) thiocyanate typically consist of sodium bicarbonate or a mixture of naphtha pitch, linseed oil, fuming nitric acid, and picric acid.
- In his 1941 book, *The Chemistry of Powder and Explosives*, Dr. Tenney L. Davis, emeritus professor of organic chemistry at the Massachusetts Institute of Technology, describes the procedures for manufacturing black, nonmercury snakes using naphtha pitch and linseed oil. George W. Weingart published a similar recipe in his 1947 book, *Pyrotechnics*, as did Takeo Shimizu in his 1981 book, *Fireworks: The Art, Science and Technique*.

Carbon Footprints

- Carbon, the sixth most abundant element in the universe, has been known since ancient times.
- Most commonly obtained from coal deposits, carbon occurs naturally as charcoal, graphite, and diamond.
- Amorphous carbon is a black soot used to make inks, paints, rubber products, and the cores of most dry-cell batteries.
- The branch of organic chemistry is devoted to the study of the nearly 10 million known carbon compounds.
- Carbon forms the bulk of all known living matter on earth.

8. Psychedelic Paper

WHAT YOU NEED

- 24 toothpicks
- 2 pieces of corrugated cardboard (2 inches wide by 8 inches long, with the fluted strip of corrugation along the length)
- Newspaper
- Can of shaving cream
- Cookie sheet
- Tempera paint (6 different colors)
- Water
- 6 paper cups
- Several sheets of white copy paper (8.5 inches by 11 inches)

WHAT TO DO

1. To create a wide-toothed comb for swirling the colors, insert several toothpicks into the fluting along one length of one piece of corrugated cardboard, approximately ½ inch apart.

2. Spread sheets of newspaper over your work area.
3. Spread a layer of shaving cream about 1 inch thick onto the cookie sheet. Make the shaving cream level by spreading it out using the second piece of corrugated cardboard as a scraper.
4. Use water to thin each color of tempera paint in its own paper cup.
5. Pour different shapes or patterns on top of the layer of shaving cream.
6. Using the toothpick comb, swirl the paint on top of the shaving cream (without pushing it deep into the shaving cream).
7. Place a piece of white paper on top of the design and press down lightly. Remove the paper from the foam, place the foam side up on a table, and use the second piece of cardboard to squeegee off the excess shaving cream.
8. Place the sheet of paper with the psychedelic design face up on a sheet of newspaper and let dry.
9. Reuse the layer of shaving cream in the cookie sheet several times to print more psychedelic paper, and refresh with shaving cream and paint when necessary.

WHAT HAPPENS

The swirling designs created in the shaving cream adhere to the paper, creating marbled paper.

WHY IT WORKS

The molecules of tempera paint, water, shaving cream, and the surface of the paper all chemically combine, causing a chemical bond that makes the swirling colors stick to the paper. The shaving cream (a chemical compound called triethanolammonium stearate) has a hydrophilic head (composed of a carboxylate ion and triethaneammonium ion) and a hydrophobic tail (made of a 17-carbon-long aliphatic chain from stearic acid). In other words, the tops of the molecules attract water and the bottoms of the molecules repel water. The water-based tempera paint is simultaneously attracted to the hydrophilic head of the shaving cream and repelled by the hydrophobic tail, which keeps the paint suspended on top of the shaving cream. When you place the paper on top of the shaving cream, the absorbent cellulose captures the paint.

WACKY FACTS

- The ionic chemical compound triethanolammonium stearate, created by the reaction between the base triethanolamine and the acid stearic acid, is composed solely of carbon, hydrogen, and oxygen.
- People marbled paper in Japan and China as early as the 12th century. According to a Japanese legend, the gods gave knowledge of the marbling process to a man named Jiyemon Hiroba in 1151 CE as a reward for his devotion to the Kasuga Shrine in the city of Nara, the first permanent capital of Japan.
- For centuries, paper- marbling artisans worked in secrecy to maintain a shroud of mystery to prevent others from mastering the craft and going into business for themselves.
- In eastern Asia, the earliest existing sheet of marbled paper originated in Turkey and dates from 1447 CE. In Turkish-style marbling, the artisan coats one side of the paper with alum to help the color adhere to the paper. Colored ink or paint is sprinkled on the surface of a thick liquid blended from gelatin (made from carrageenan, an ingredient extracted from seaweed) and water and poured into a shallow tray.
- In Europe, paper marbling flourished in Italy as early as the 16th century, and today the craft survives almost exclusively in Florence.

IT'S ELEMENTARY

Pardon the Argon

In 1785, British scientist Henry Cavendish discovered that air is composed of 99.3 percent nitrogen, oxygen, and carbon dioxide and 0.7 percent an inert gas that he could not identify. More than a century later, in 1894, English physicist John William Strutt (better known as Lord Rayleigh) and Scottish chemist Sir William Ramsey used spectroscopy to identify the colorless, odorless, and tasteless gas, which would not react with anything else. They named the element argon, derived from the Greek word *argos*, meaning "lazy."

The decay of radioactive potassium-40 produces argon, slowly increasing the presence of argon in the atmosphere. Because argon does not react with other elements, manufacturers use the gas in industrial processes that require a nonreactive atmosphere, such as producing semiconductor wafers. Incandescent lightbulbs are filled with argon to prevent oxygen from corroding the hot filament.

Window manufacturers use argon to fill the space between the panes of double glass because the gas is a good insulator. The US National Archives minimizes the deterioration of a copy of the Magna Carta by storing the parchment document in a display box filled with argon gas, which, unlike oxygen, does not degrade fragile documents.

NITROGEN

14.0067 2 5
N
7

9. Hilarious Hot-Air Balloon

WHAT YOU NEED

- Scissors
- 6 bendable plastic drinking straws
- Scotch Tape
- Ruler
- Spool of thread
- 4-by-4-inch square of aluminum foil
- Indelible marker
- 4 birthday candles
- Butane barbecue lighter
- 1 tall, clear kitchen trash bag (16 gallon, 24 inch by 33 inch by 0.31 mil)

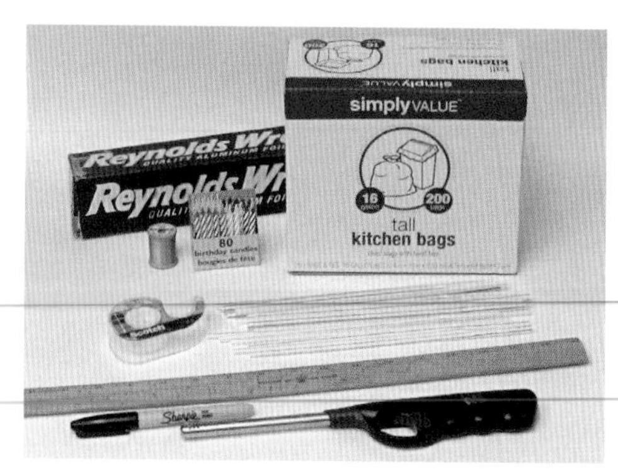

WHAT TO DO

1. Using the scissors, snip off the shorter end of the bendable straw that includes the accordion-type hinge. Snip 2-inch slits in one end of a straw, insert the slit end into one end of a second straw, and secure in place with a small strip of Scotch Tape. Repeat until you make a length of straw 17 inches long (or 70 percent of the width of the plastic trash bag). Repeat this process to create a second, matching length of straw.
2. Cross the two lengths of straw to create an *X*, and secure the two straw lengths in place with a piece of Scotch Tape, as shown.

3. Tie the end of the thread from the spool to the spot where the straws cross. The thread will serve as a leash (or tether) for the hot-air balloon.
4. Using the ruler, measure 1 inch in from the midpoint of each side of the square of aluminum foil and make a dot with the indelible marker for a total of four dots. Use the butane barbecue lighter to melt two drops of hot wax from the end of one birthday candle onto one of the dots on the aluminum foil, and stand the candle upright in the hot wax, holding it in place until the wax hardens. Repeat this process to secure each of the remaining three candles on the remaining three dots.

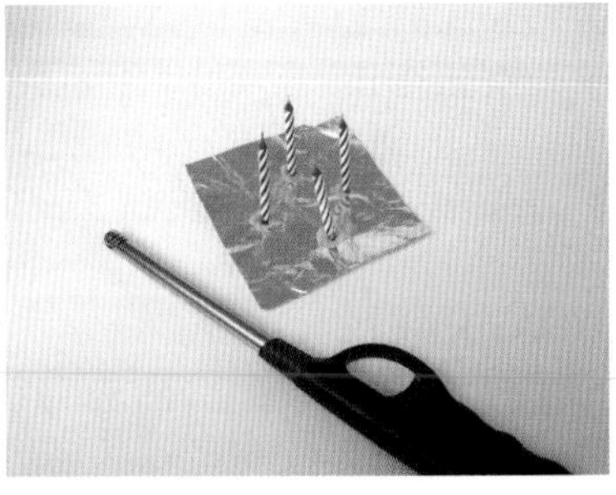

5. Place the aluminum candle platform upright in the center of the straw crossbeam so that the corners of the aluminum foil square touch the straws, and tape it in place to the straws from underneath the aluminum foil. (Make sure the spots where the candles are

mounted to the aluminum foil do not sit on top of the straws; otherwise the heat from the candle flames will melt the straws.)

6. Open the plastic trash bag and, using Scotch Tape, carefully attach the four ends of the straws to the inside edges of the rim of the trash bag, so the candles face inside the trash bag.
7. Have a partner hold the bottom of the trash bag 3 feet off the ground, and use the butane barbecue lighter to carefully ignite the four candles without letting the flames touch or get near the sides of the plastic bag. Hold out the sides of the plastic bag so the hot air produced by the candles does not melt the plastic together. Allow the plastic bag to fill with hot air and then release the bag, holding the spool of thread as a tether line.

8. Keep a bucket of water or a fire extinguisher nearby, and do not stand directly under the hot-air balloon in case wax from the candles drips from the balloon. If outside, do not release the tether to float the balloon freely.
9. Should a lit candle accidentally fall off the aluminum foil, the drop through the air will, in most cases, extinguish the flame. The sides of the plastic bag, if touched by flame, will melt without catching fire; however, be cautious of droplets of hot plastic.

WHAT HAPPENS

As hot air rises inside the plastic bag, the bag fully inflates and rises. The plastic bag drifts upward until the flame goes out and the air inside the plastic bag cools. If the balloon is launched at night, the flame also illuminates the balloon.

WHY IT WORKS

When air (composed of 78 percent nitrogen, 21 percent oxygen, and small amounts of other gases) is heated, the molecules move faster, causing them to expand and become less dense, increasing the volume of the gas. In other words, cold air is denser than hot air. The denser, cooler air pushes the warmer, less dense air upward. The hot air inside the plastic bag, lighter than the surrounding cool air, causes the plastic bag to rise.

WACKY FACTS

- Never launch a hot-air balloon outdoors without a tether. Slight breezes and air currents can send an unpredictable and difficult-to-control hot-air balloon sailing into a tree, wooded area, or field or cause it to land on a wooden building, inadvertently starting an uncontrolled fire and causing a tragic event. A candle-powered hot-air balloon can travel up to a mile or more depending on the local winds and atmospheric conditions. Anyone who launches a candle-powered hot-air balloon outdoors can be held financially responsible for any damages caused.

- For centuries in countries across Asia, people have launched multitudes of sky lanterns—small hot-air balloons made of paper—for good luck and sent wishes skyward during traditional festivals, including the Lanna Yi Peng festival (in Chiang Mai, Thailand) and the Lantern Festival (in China, Taiwan, and Vietnam).
- On July 3, 2011, a sky lantern in Myrtle Beach, South Carolina, started a fire that consumed 805 acres.
- On July 11, 2011, a lit Chinese paper lantern landed on top of a timber-frame farm home in Trowbridge, England, igniting the insulation in the roof and forcing a family of three to flee the abode.
- On July 1, 2013, a lit paper lantern set fire to a plastics recycling plant in Smethwick, England.
- In Sanya, the Chinese city where paper lanterns were first created, the government banned sky lanterns after an abundance of lanterns hovering in the sky during a festival delayed flights at a nearby airport.
- In the United States, more than two dozen states and many local jurisdictions have either banned sky lanterns by law or by regulation because the devices pose serious fire and safety hazards.
- The Boy Scouts of America has deemed that the release of a sky lantern conflicts with fundamental scouting safety principles that relate to fire management—because a person cannot attend to the "recreational fire" at all times until it is completely extinguished.

A Breath of Fresh Nitrogen

- Scottish chemist Daniel Rutherford first isolated nitrogen in 1772. He called the gas "noxious air."
- In 1790, French chemist Jean-Antoine Chaptal named nitrogen after the mineral *niter* when he discovered that niter contained the gas. Niter is more commonly known as saltpeter or potassium nitrate.
- Although we often refer to the air we breathe as "oxygen," the air we breathe is 78 percent nitrogen gas.
- Many explosives contain nitrogen, most notably TNT, nitroglycerin, and gunpowder.
- Triton, the largest of Neptune's 13 moons, has geysers that spew plumes of nitrogen five miles high. The temperature on Triton is so cold that nitrogen sits on the moon's surface as rock-hard frost.

10. Smoky Cold Pack

SAFETY FIRST

This experiment should be performed outside.

- Safety goggles
- Rubber gloves

WHAT YOU NEED

- Scissors
- 1 instant cold pack (made with ammonium nitrate)
- Bucket
- Water
- Newspaper
- Rubber bands (or string)
- Butane lighter or matches

WHAT TO DO

1. Wearing safety goggles and a pair of rubber gloves, use a pair of scissors to carefully cut open the instant cold pack. Inside you'll discover white granules of ammonium nitrate and a small plastic bag filled with water. (Do not ingest any ammonium nitrate, and avoid contact with skin.)
2. Remove and discard the bag of water.
3. Pour the granules into the bucket.
4. Slowly add water and gently swirl the bucket until you have added just enough water to allow the ammonium nitrate granules to dissolve, creating roughly 1 inch of solution.
5. Fold a sheet of newspaper in half several times until it will fit into the bucket. Submerge the sheet of newspaper in the liquid to saturate it in the solution for a few minutes, remove it from the bucket, and gently reopen the newspaper, lay it on a flat concrete surface (such as a driveway or sidewalk), and let dry in the sun undisturbed for a few hours. To prevent breezes from blowing the newspaper away, place rocks or weights on the four corners.
6. Repeat with additional sheets of newspaper until no more solution remains.

7. When the sheets of newspaper are completely dry, take a full sheet, fold it in half, roll it up tightly, and secure shut with a rubber band (or piece of string).
8. Using a butane lighter or matches, ignite one end of the tube of newspaper, place on the ground, and back away quickly.

WHAT HAPPENS

The tube of newspaper spews a plume of white smoke.

HOW IT WORKS

When ignited, the ammonium nitrate (a compound of nitrogen, hydrogen, and oxygen) impregnated in the newspaper produces thick plumes of white smoke.

WACKY FACTS

- An instant cold pack consists of two bags. The first bag contains ammonium nitrate and an inner bag filled with water. Squeezing the package pops the inner bag, spilling the water into the larger bag, which dissolves the ammonium nitrate, triggering an endothermic reaction. That means the hydrated ammonium nitrate absorbs heat from the air inside the bag, making the bag cold.
- When heated, ammonium nitrate decomposes into nitrous oxide gas and water vapor.
- Ammonium nitrate, a common ingredient in fertilizer, can be mixed with certain hydrocarbons to make explosive bombs.

11. Homemade Ping-Pong Ball Smoke Bomb

SAFETY FIRST

This experiment should be performed outside.

WHAT YOU NEED

- Sharp scissors
- 4 Ping-Pong balls
- Sharpened pencil
- Ruler
- Aluminum foil
- Tongs
- Butane barbecue lighter

WHAT TO DO

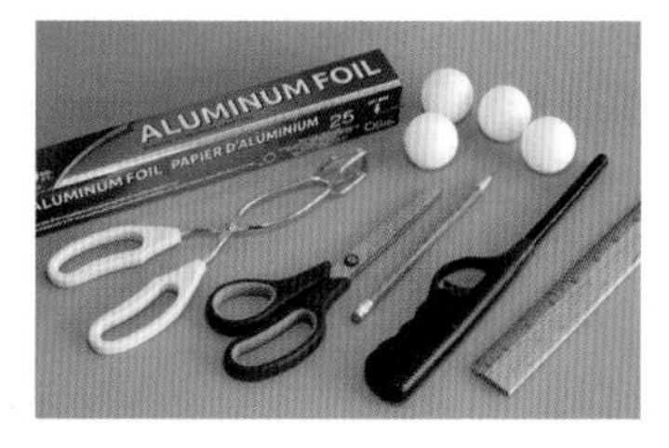

1. Carefully use the sharp tip of the scissors to poke a hole in the side of one Ping-Pong ball. Make the hole approximately the diameter of the pencil.
2. Use the scissors to cut the remaining three Ping-Pong balls into narrow strips that will fit through the hole in the first ball.

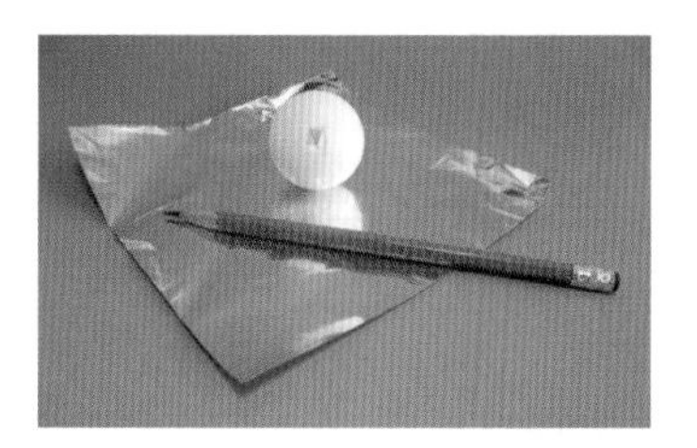

3. Insert the strips through the hole and into the first Ping-Pong ball to fill it.
4. Insert the pencil into the hole.
5. Cut a 6-inch square of aluminum foil. Wrap aluminum foil around the ball and the first inch or two of the pencil.

6. Remove the pencil, leaving a foil-wrapped ball connected to a short foil tube.

7. Outside in a well-ventilated area, use the tongs to hold the smoke bomb by the foil tube (without crushing the tube), and use the butane barbecue lighter flame to heat the aluminum foil on the bottom of the Ping-Pong ball. The moment smoke starts to fume from the nozzle, set the smoke bomb on the ground, step back, and observe. **Do not breathe the smoke**, which is toxic.
8. When the smoke bomb exhausts itself, wait 5 minutes for the aluminum foil to cool down before touching it with your bare hands. Discard the foil.

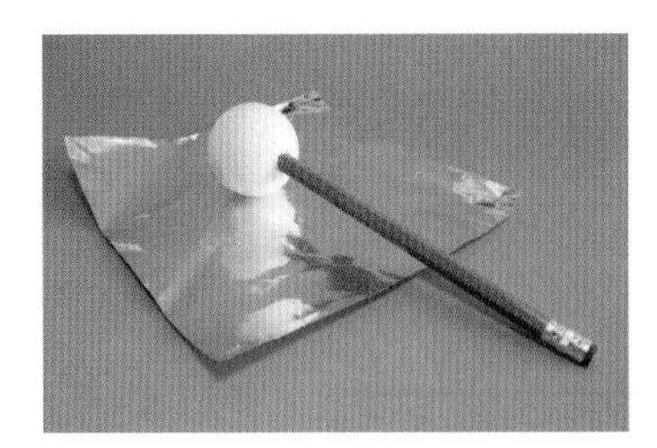

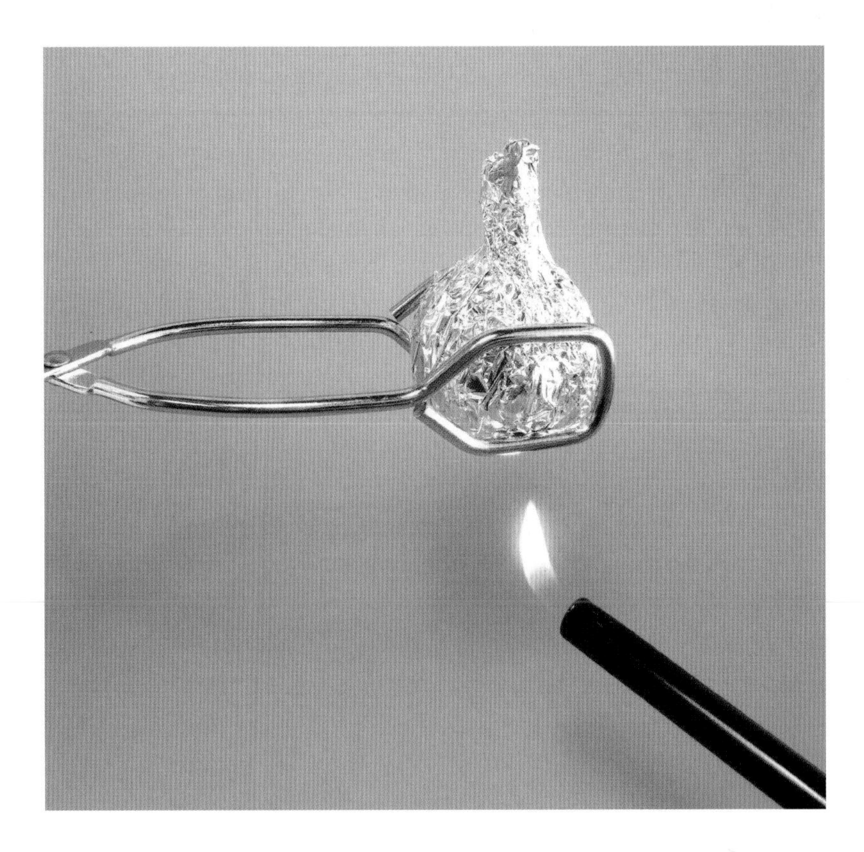

WHAT HAPPENS

The aluminum foil–wrapped Ping-Pong ball emits a plume of white smoke for roughly 15 seconds.

WHY IT WORKS

Ping-Pong balls are made of celluloid, a highly flammable plastic made from camphor and nitrocellulose. Most nitrocellulose contains from 10.5 to 13.5 percent nitrogen. The degree of nitration determines the flammability of the final product. A celluloid ball burned in the open produces some smoke. However, wrapping the ball to control the amount of oxygen available slows the rate of combustion, creating more smoke, and the spout regulates the dispersal of the smoke.

WACKY FACTS

- Early table tennis balls, consisting of acidified celluloid, would sometimes combust or explode when hit. Acidified celluloid grew increasingly unstable over time, and the heat from friction could ignite it.
- In 1846, Christian Schönbein invented nitrocellulose by accident. Experimenting in his kitchen with nitric and sulfuric acid, he broke a bottle, wiped up the mess with a cotton apron, and put it on the stove to dry. It burst into flames—explosively, with four times the power of gunpowder. Schönbein dubbed the new smokeless explosive *guncotton*.
- Nitrocellulose is the main ingredient in modern gunpowder.
- Inventor and industrialist John Wesley Hyatt mixed solid nitrocellulose, camphor, and alcohol under pressure to create a plastic that he and his brother Isaiah patented and registered under the trade name Celluloid in 1873. The compound was also used to make billiard balls, false teeth, knife and brush handles, combs, piano keys, dice, buttons, fountain pens, and celluloid collars for men's shirts.
- Motion picture film was initially made of celluloid, which often caught fire and decomposed quickly. The movie industry replaced celluloid film with more-stable plastic films, including cellulose acetate and polyethylene.

OXYGEN

15.9994 2 6
O
8

12. Freaky Soap Soufflé

WHAT YOU NEED

- Bar of soap (Ivory works best)
- Dish, microwave safe
- Microwave oven

WHAT TO DO

1. Place the bar of soap on a microwave-safe dish in the microwave oven.
2. Heat for 2 minutes.
3. Let sit for 3 minutes to cool.
4. Open the microwave oven and remove the soap soufflé.
5. Do not discard the altered soap, which can still be used for its original purpose.

WHAT HAPPENS

The bar of soap transforms into a large, lava-like formation resembling a flaky cloud or massive foam sculpture, more than six times the size of the original bar.

WHY IT WORKS

The heat from the microwave oven vaporizes the water (a compound of hydrogen and oxygen) in the bar of soap and causes the tiny pockets of air (78 percent nitrogen, 21 percent oxygen, and trace elements) trapped in the soap to expand in all directions, pushing the soap outward, transmogrifying into a bizarre, light, spongy sculpture of sorts. When the soap cools, it stiffens but retains its new shape.

WACKY FACTS

- The reaction tends to infuse the microwave oven and kitchen with the scent of soap, in some cases refreshing the air in the house.
- Ivory soap works best for this experiment because the manufacturer whips air into the soap, resulting in a bar of soap that floats in water.
- When any substance is heated, its molecules and atoms vibrate faster, increasing the amount of space between the atoms.
- Popcorn kernels pop in a microwave oven because the water molecules inside the kernel turn to steam as the kernel heats up. The steam increases the pressure inside the kernel, causing the shell of the kernel to explode and the meat of the kernel to expand.
- A bar of soap heated in a pot on a stove melts into liquid soap. Unlike a microwave oven, which generates microwaves that cause soap and trapped air molecules to vibrate, producing heat, a stove heats by convection. In other words, a burner on a stove heats the bottom of the pot and the soap sitting closest to it, making the bottom of the bar of soap hotter than the rest of the bar.

13. Daffy Dry Ice Monster Bubble

SAFETY FIRST

Do not touch dry ice with your bare hands.

- Work gloves

WHAT YOU NEED

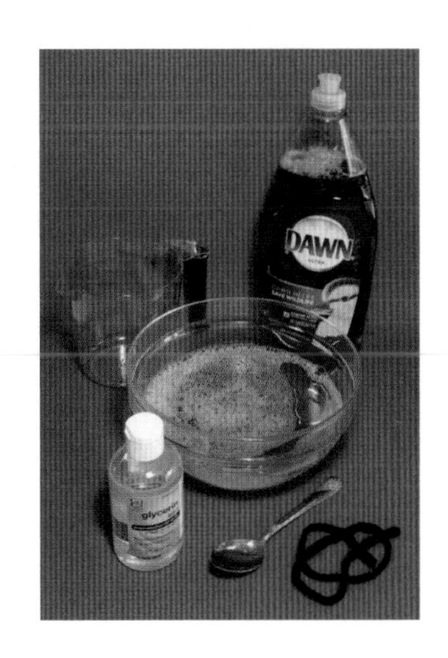

- Measuring cup
- Dishwashing liquid (Dawn works best)
- Glycerin (available at drugstores)
- Water
- 1-quart bowl
- Large glass mixing bowl with a lip around the rim
- Warm water
- Dry ice (available at many grocery stores or ice-cream shops)
- 18-inch length of yarn

WHAT TO DO

1. Mix 2 ounces of dishwashing liquid, ½ ounce of glycerin, and 2½ cups of warm water in the 1-quart bowl to make a bubble solution.
2. Fill the large glass mixing bowl halfway with warm water.
3. Wearing work gloves (to avoid touching the dry ice with your bare hands), gently place the dry ice in the glass mixing bowl. (Dry ice can burn your skin if it touches it directly. Also never eat dry ice, as it can be fatal.) The water will start bubbling and emitting smoke (like a witch's cauldron). Avoid breathing the vapor, which is carbon dioxide.

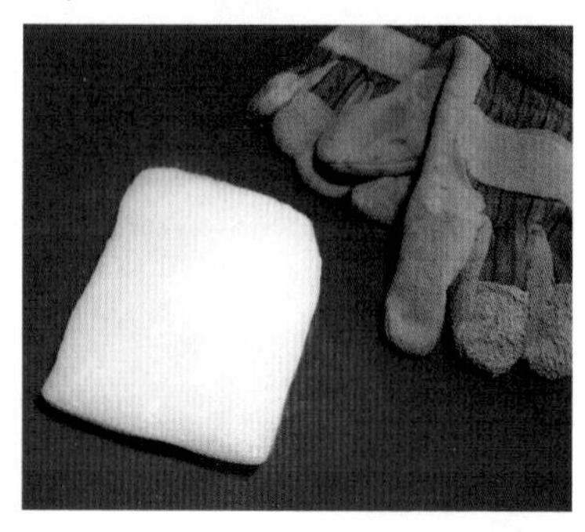

4. Dip your fingertip into the bubble solution and run it around the lip of the bowl to wet the lip, without getting any of the bubble mixture in the bowl.
5. Saturate the length of yarn in the bubble solution, and holding the two ends of the yarn to keep the string taut, slide the saturated length of yarn across the top of the bowl to form a bubble film across the rim.
6. Observe what happens.

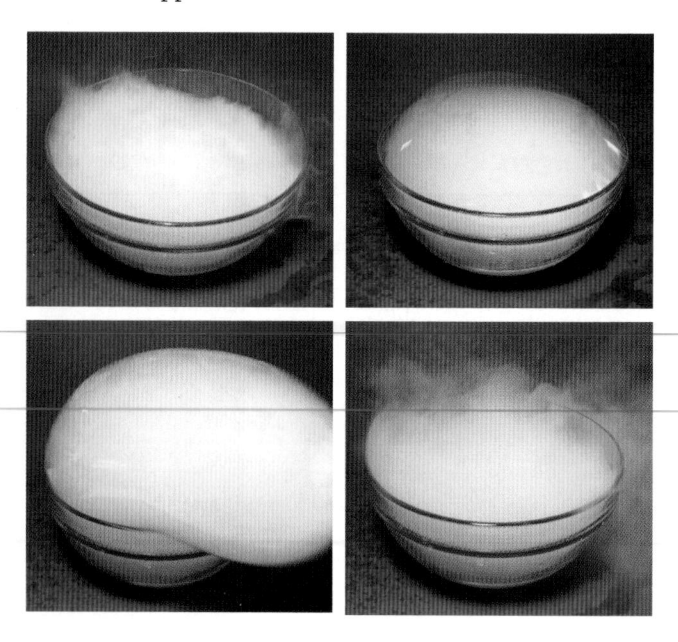

WHAT HAPPENS

A giant soap bubble forms on the rim of the bowl, fills with smoke, and eventually collapses and pops, spreading smoky fog across the countertop.

WHY IT WORKS

At room temperature, dry ice (carbon dioxide frozen into its solid form) sublimes, changing directly from a solid to a gas, bypassing the liquid stage. The warm water accelerates sublimation, generating carbon dioxide gas (a compound of carbon and oxygen), which fills the bubble. Eventually, excess pressure causes the bubble to explode, releasing the carbon dioxide gas.

WACKY FACTS

- In 2014, scientists at the University of California, Davis, developed an instrument to split a molecule of carbon dioxide into carbon and oxygen, according to a study published in the journal *Science*.
- To generate small smoke-filled bubbles, simply place a chunk of dry ice in a bowl of the bubble solution. The dry ice will continue generating bubbles until it sublimes completely or the bubble solution is exhausted, whichever comes first.

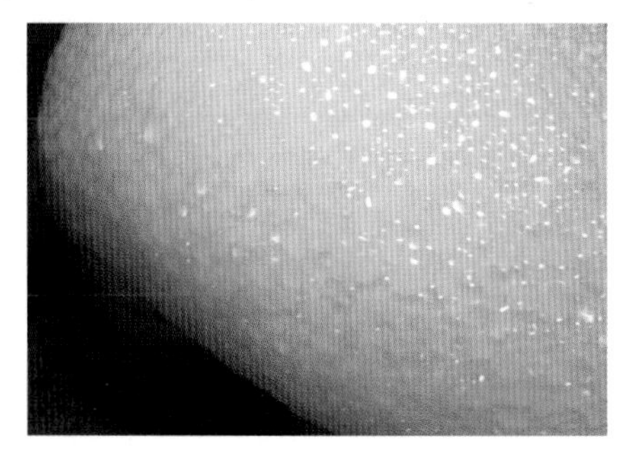

- Dry ice is colder than ice made from frozen water. Dry ice can reach a temperature as low as −112°F.
- To make dry ice, carbon dioxide gas is compressed into a liquid and then cooled and evaporated to make carbon dioxide snow, which is then compressed into blocks of solid dry ice.
- If ingested, dry ice can cause death due to its low temperature.
- Dry ice can be used to ship perishable foods because the ice does not melt.
- Theatrical productions often use dry ice to create a dense, foggy effect.
- Inserting a small piece of dry ice into an uninflated balloon and then tying a knot in the neck of the balloon will produce enough carbon dioxide to inflate the balloon.

Oxygen: Can't Live Without It

- Oxygen is the third most abundant element in the universe.
- Oxygen is the most abundant element in the human body, comprising approximately 65 percent of the body's mass.
- In the 1770s, French scientist Antoine Laurent Lavoisier, convinced that his newly discovered gas was the active ingredient in all acids, inaccurately named the gas oxygen (after the Greek word *oxygenes*, meaning "acid maker"). Oxygen is not essential to acidic properties.
- In 1771, Swedish chemist Carl Wilhelm Scheele heated mercuric oxide and discovered that the resulting gas was one of two gasses contained in air. Scheele wrote a book titled *A Chemical Treatise on Air and Fire* describing his experiments and asked his patron, Swedish chemist Torbern Bergman, to write the introduction. Bergman took so long to write the introduction that the publisher was unable to print the book until 1777, by which time English chemist Joseph Priestly had reported his own experiments and had taken credit for the discovery of oxygen.
- On Earth, the oxygen in the atmosphere is produced by plants through photosynthesis. Without plants, the air would contain very little oxygen.

14. Cornstarch-Powered Flamethrower

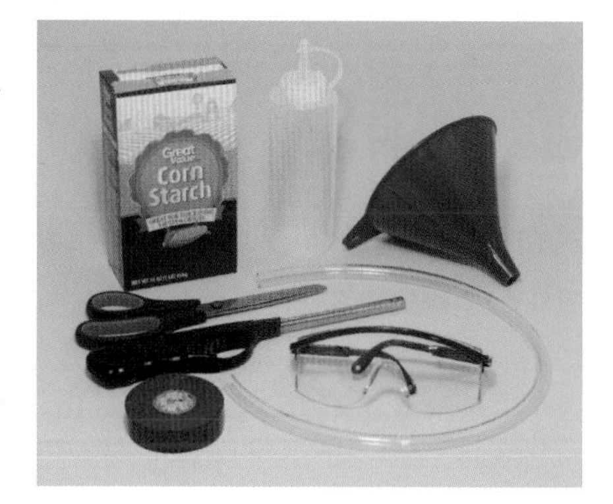

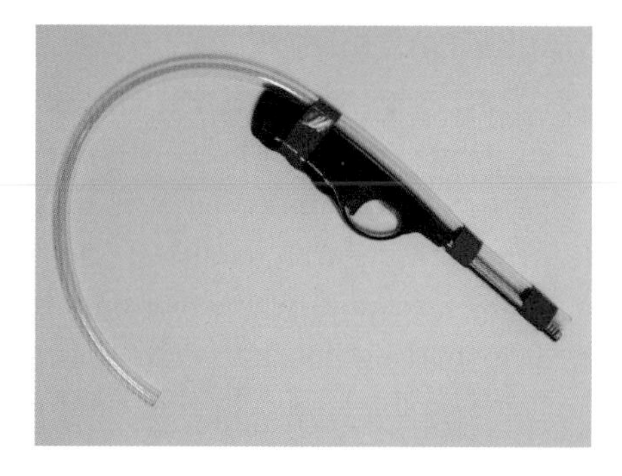

SAFETY FIRST

This experiment should be performed outside.

- Safety goggles
- Fire extinguisher

WHAT YOU NEED

- Clear plastic tube, ¼-inch diameter, 24 inches in length (available at hardware stores or aquarium supply stores)
- Butane barbecue lighter
- Electrical tape
- Scissors
- Funnel
- Plastic squeeze bottle with dispenser tip
- Cornstarch

WHAT TO DO

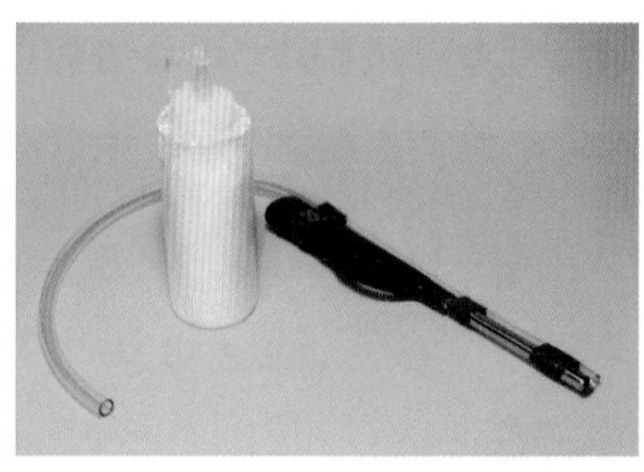

1. Place the end of the clear plastic tube alongside and underneath the tip of the butane barbecue lighter so that the opening of the tube aligns with the opening of the lighter.
2. Use electrical tape to secure the end of the tube to the lighter—as if creating a double-barreled shotgun. Use the scissors to trim the length of the electrical tape needed.
3. Using the funnel, fill the squeeze bottle with cornstarch.
4. Insert the tip of the squeeze bottle into the end of the plastic tube taped to the lighter, and squeeze cornstarch to fill the tube.
5. Outside, wearing safety goggles and with a fire extinguisher on hand, place the free end of the tube in your mouth.
6. Holding the lighter at arm's distance and pointing the end away from any people, animals, plants, or flammable objects, ignite the lighter.
7. With the flame burning, blow forcefully through the free end of the tube.

WHAT HAPPENS

A long flame blasts from the butane lighter, creating a fireball.

WHY IT WORKS

Cornstarch, when dispersed in air, is flammable due to the oxygen around it. The fire cannot jump back down the tube because the cornstarch must be dispersed in air to burn.

WACKY FACTS

- Cornstarch contains starch, chains of glucose molecules, which, when suspended in air, are extremely flammable. When airborne cornstarch particles surrounded by oxygen ignite, they can quickly set alight nearby cornstarch particles, one after the other, causing an explosion—explaining why explosions often occur in grain-storage facilities.
- Transporting flour raises the risk of explosion because the smallest flame or spark in a cloud of flour particles suspended in the air can trigger an explosion.

15. Loony Lava Lamp

WHAT YOU NEED

- Funnel
- Measuring cup
- Water
- 1-liter plastic soda bottle, clean and empty
- Vegetable oil
- Food coloring (red, blue, or green)
- Alka-Seltzer tablets
- Lantern flashlight

WHAT TO DO

1. Using the funnel, pour ¾ cup of water into the clean, empty 1-liter plastic soda bottle.
2. Still using the funnel, fill the rest of the soda bottle with vegetable oil, leaving 1 inch of air below the neck of the bottle.
3. Let the liquids sit until the oil and water separate completely into two layers.
4. Add 10 drops of either red, blue, or green food coloring to the mixture in the bottle. Observe as the drops of food coloring sink through the layer of oil, come to rest on the surface of the water, and after a few minutes, burst through to the water.

5. Crack 3 Alka-Seltzer tablets in half and drop them into the bottle.
6. Turn on the lantern flashlight, stand it upright, and carefully place the bottle on the plastic faceplate covering the flashlight's lightbulb.
7. Observe.
8. When the action ceases, you can add more broken tablets of Alka-Seltzer or cap the bottle tightly and save the lava lamp for future use.

WHAT HAPPENS

A continuous stream of colored bubbles rises to the top of the bottle and slowly sinks back down to the bottom again.

WHY IT WORKS

The water and oil are immiscible, meaning they do not mix together. The water falls to the bottom because it is denser than the oil. The food coloring, being water soluble, sinks to the bottom, breaks through the surface tension of the water, and mixes with the water. The Alka-Seltzer reacts with the water, creating bubbles of carbon dioxide gas (a compound of carbon and oxygen), which, being less dense than both the water and the oil, rise to the surface, pushing globules of colored water up with them. At the surface, the carbon dioxide escapes, and globules of colored water sink back down through the oil to the layer of water.

WACKY FACTS

- Craven Walker, a native of Singapore, came up with the idea for the lava lamp while drinking in a pub in Dorset, England, after World War II and spent the next 15 years developing the lamp. Walker launched the "Astro lamp" in 1963, just in time for the psychedelic 1960s.
- Carbon dioxide, a chemical molecule consisting of one carbon atom covalently bonded to two oxygen atoms, is a colorless, odorless, incombustible gas resulting from the oxidation of carbon.
- Manufacturers use carbon dioxide to carbonate soft drinks. When heated, baking powder and baking soda release carbon dioxide, which leavens bread. Baker's yeast ferments the sugars within dough, producing carbon dioxide.
- Fire extinguishers frequently contain carbon dioxide, which extinguishes electrical fires.
- Carbon dioxide is a greenhouse gas produced by human activities, primarily by burning fossil fuels.

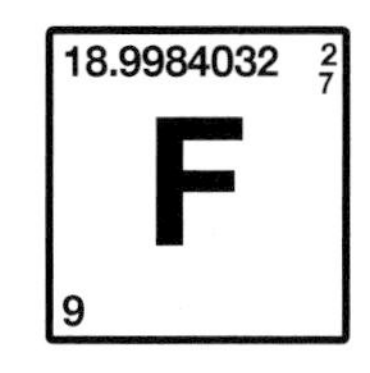

FLUORINE

16. Eggciting Eggsperiment

WHAT YOU NEED

- 1 bottle of fluoride rinse containing 0.05 percent sodium fluoride (such as ACT Anticavity Fluoride Rinse or Colgate Phos-Flur Rinse)
- 2 hard-boiled eggs
- Paper towel
- Cola
- 3 drinking glasses
- Water

WHAT TO DO

1. Pour 4 inches of fluoride rinse into one of the drinking glasses.

2. Gently place one hard-boiled egg in the solution, and let it sit undisturbed for 5 minutes.
3. Remove the egg from the fluoride rinse solution, discard the solution, and gently pat the egg dry with a sheet of paper towel.
4. Pour 4 inches of cola into the two remaining glasses.
5. Place the treated egg in one glass of cola and the untreated egg in the other cola. Replace the cola every 12 hours for 2 days, noting the changes in the two eggs.

6. Remove the eggs from the drinking glasses, rinse the eggs clean with water, and observe the changes in the eggs.

WHAT HAPPENS

The cola stains and etches into the shell of the untreated egg. The shell of the egg treated with the fluoride solution reacts much more slowly to the cola.

WHY IT WORKS

By soaking the egg in the fluoride solution, the fluoride (a negatively charged ion of fluorine) integrates with the calcium in the eggshell, strengthening the shell and protecting it from the acids in the cola.

WACKY FACTS

- Like the fluoride solution with the egg, the fluoride in toothpaste, mouthwashes, and professional treatments protects tooth enamel from decay caused by acids in foods, beverages, and bacteria (which naturally occur in the mouth).
- To protect your teeth from decay, brush your teeth twice daily for two minutes with fluoridated toothpaste and drink fluoridated water or milk, rather than acidic soda or sugary drinks.

- According to the American Dental Association, studies prove that fluoridation of community water supplies in the United States effectively reduces dental decay by at least 25 percent, despite the widespread availability of fluoride from other sources, such as fluoride toothpastes and mouthwashes.
- The American Dental Association estimates that for most cities, every dollar invested in water fluoridation saves $38 in dental treatment costs.
- As of 2012, approximately 75 percent of the population of the United States is served by fluoridated community-water systems.

Fluorine: The Lean, Mean, Cavity-Fighting Machine

- French scientist André-Marie Ampère coined the name *fluorine* in 1812, derived from the mineral fluorite, named from the Latin word *fluere*, meaning "to flow."
- In 1886, French chemist Joseph Henri Moissan first isolated fluorine, after his work was interrupted four times by serious poisoning caused by the element he was pursuing.
- The gemstone topaz contains fluorine.
- Fluorine, a greenish-yellow gas, combines with other elements more readily than any other chemical element, forming compounds called fluorides.
- Compounds of fluorine are used in refrigerator cooling systems, aerosol cans, the preparation of uranium for atomic bombs, toothpastes, and public drinking-water supplies (to prevent tooth decay).

NEON

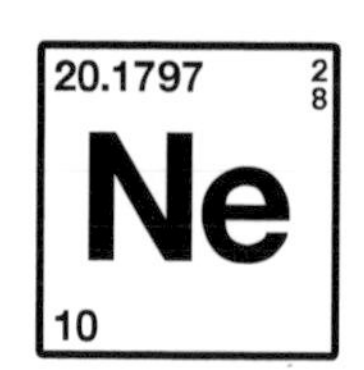

17. Surprising Static Bulb

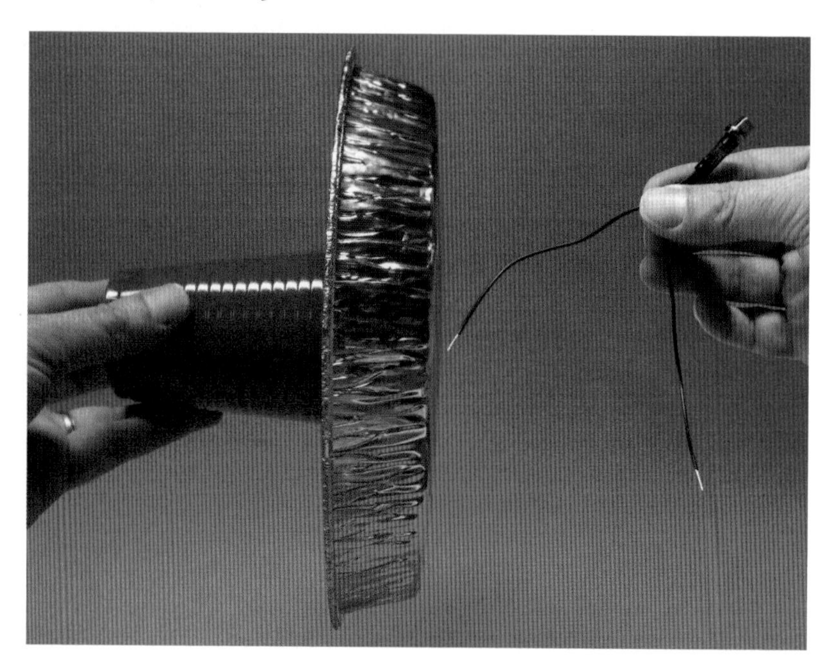

WHAT YOU NEED

- Neon bulb (NE-2—available at RadioShack)
- Aluminum pie plate
- Hot glue gun
- Plastic drinking cup
- Plastic cutting board
- Wool scarf or sweater

WHAT TO DO

1. Bend the two wires attached to the neon bulb at right angles (perpendicular) to the bulb.
2. If necessary, use your fingers to burnish the bottom of the

aluminum pie plate to flatten any indentations and make the bottom smooth.

3. Carefully use the hot glue gun to attach the rim of the plastic drinking cup to the center of the aluminum pie plate. The plastic cup will serve as an insulating handle to lift and hold the pie plate.

4. Place the plastic cutting board on a table or countertop, and rub a wool scarf or sweater over the cutting board for 30 seconds, giving the plastic board an electrical charge.
5. Holding the plastic cup as a handle, place the aluminum pie plate on the plastic cutting board.
6. Lift the handle to raise the pie plate from the cutting board.
7. Touch the tip of your finger to the rim of the plate.
8. Turn out the lights and touch one of the wires attached to the neon bulb to the bottom of the aluminum pie plate. (Results will vary based on the temperature and humidity.)

WHAT HAPPENS

When you touch the tip of your finger to the aluminum pie plate, you will feel a slight spark of electricity. When you touch one of the wires of the NE-2 neon bulb to the bottom of the plate, the bulb flashes bright light.

WHY IT WORKS

Rubbing the plastic cutting board with the wool scarf or sweater gives the plastic board a negative electrical charge. (The plastic has a greater electron affinity than the wool, attracting electrons away from the atoms of the wool.) When the aluminum pie plate is pressed against the cutting board, the negative charges in the plastic repel the negative charges in the aluminum plate, causing them to move from the bottom surface of the aluminum plate to the rim of the aluminum plate, leaving the bottom of the aluminum plate with an excess of positive charges.

When you touch the aluminum plate, the negative electrons jump from the aluminum plate to the ground, leaving the aluminum plate with an overall positive charge distributed throughout the object.

When you touch one of the wires of the NE-2 neon bulb to the bottom of the plate, the positive charge of electrons strikes the atoms of neon gas inside the bulb, exciting the electrons in the gas and promoting them to an orbital of higher energy. When the electrons slow back down to their original orbital, a particle of light (a photon) carries away the energy of excitation and the bulb glows.

WACKY FACTS

- The plastic cutting board, wool sweater, and aluminum pie pan serve as an electrophorus, a device used for repeatedly generating static electricity by induction.
- You can use the neon bulb to determine whether an electrical charge is positive or negative. The light inside the neon bulb occurs nearest the negatively charged electrode (known as the cathode). If you touch the bottom of the charged aluminum plate with the other wire, the side of the bulb that lights up will switch.
- The static electricity generated by rubbing a wool sweater against a plastic cutting board can exceed 10,000 volts but with very low amperage, rendering the static electricity harmless—but ample to light a neon bulb (requiring 80 volts and as little as 10^{-12} amps).
- The electrophorus in this experiment requires low humidity to work successfully. Conduct this experiment on a dry day or inside an air-conditioned building.
- You can also illuminate the neon bulb on a cold day by holding one of the wires and walking across the carpet in your home, dragging your feet. This action generates static electricity that discharges through the lightbulb in your hand.

See the Light with Neon

- British chemists Sir William Ramsay and Morris W. Travers discovered neon in the atmosphere in 1898 and named the gas for the Greek word *neon*, meaning "new."
- Primarily used for filling lamps and luminous sign tubes, neon gas glows a fiery-red color when a current of electricity flows through it. Adding a few drops of mercury inside the lamp or tube turns the light a brilliant blue.
- Neon light penetrates fog. Pilots have reported seeing neon beacons for 20 miles through fog, when other lights were impossible to see.

SODIUM

22.98976928	2 8 1
Na	
11	

18. Electric Pickle

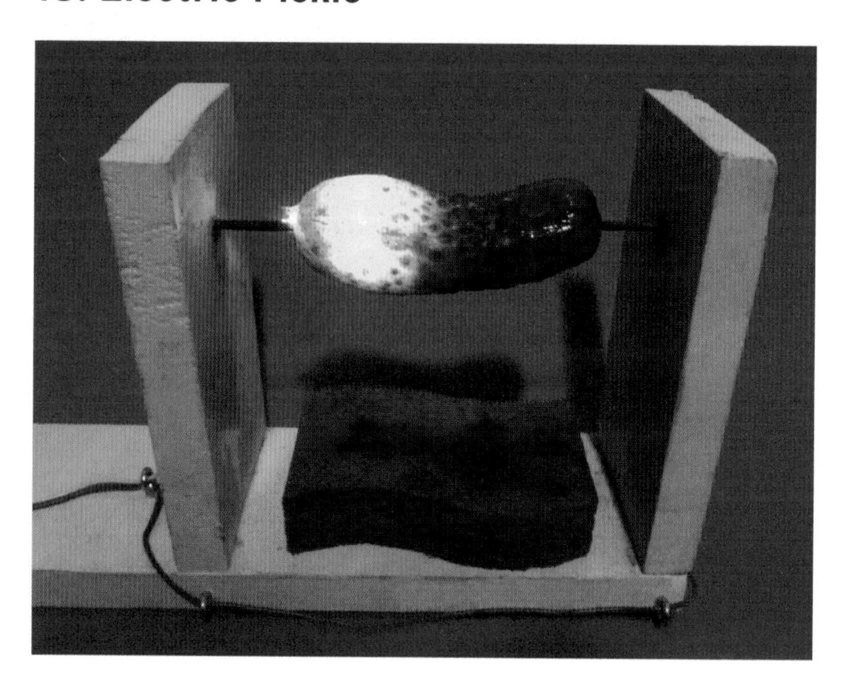

SAFETY FIRST

- Safety goggles
- Rubber-soled shoes

WHAT YOU NEED

- Ruler
- Pencil
- 2 planks of ¾-inch-thick wood (3½ by 6 inches)
- Drill with ⅛-inch bit and a screwdriver bit
- Plank of ¾-inch-thick wood (3½ by 17 inches)
- 4 wood screws, 2 inches long
- Colorful latex paint (optional)
- Paintbrush (optional)
- 5 eyelet screws
- Screwdriver
- Hammer

- Junction box to accommodate 2 light switches
- 2 wood screws, ¾ inch long
- 2 clamp combination connectors, ⅜ inch
- Pliers
- 2 screws, 3 inches long (⅛-inch diameter)
- Wire cutters
- 26-inch-long 20-gauge wire
- 14-inch-long 20-gauge wire
- Small piece of folded foam or cardboard

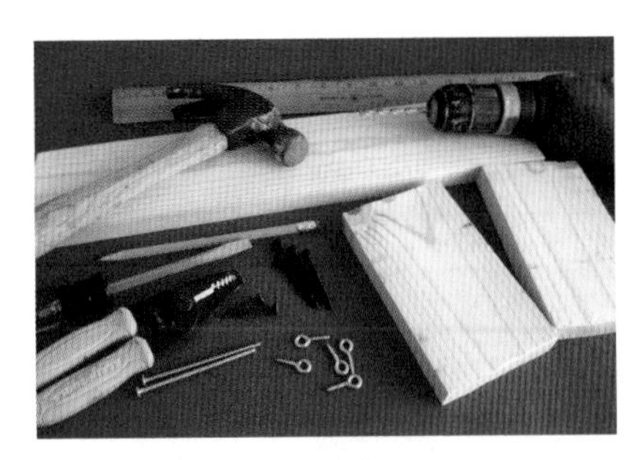

- 2 on/off light switches
- 6-foot extension cord
- Plastic faceplate for 2 light switches
- Sponge
- Large dill pickle

WHAT TO DO

1. Mark a line down the center of the length of one of the 3½-inch-by-6-inch pieces of wood, and then make a small mark 1½ inches from one end of the wood. Wearing safety goggles, drill a ⅛-inch hole through the wood at that spot. Repeat for the other 6-inch piece of wood.
2. Drill two pilot holes on one end of the piece of 3½-inch-by-17-inch lumber, then use two 2-inch wood screws to attach one of the 6-inch pieces of wood perpendicular to the end, making sure the previously drilled center hole is away from the base.
3. Screw the second piece of 6-inch wood to the piece of lumber, 6 inches away from the first

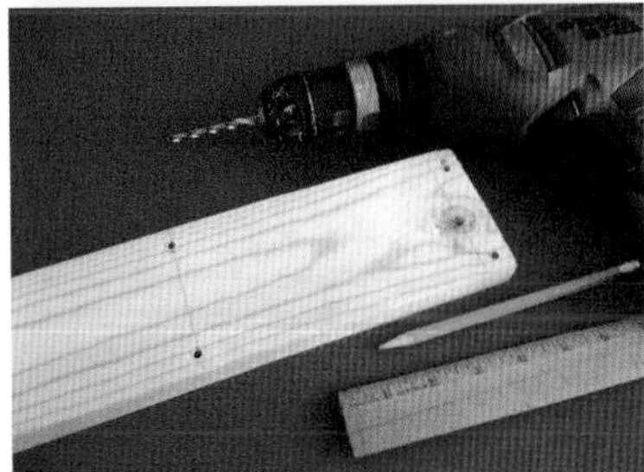

piece, again with the center hole away from the base. If desired, paint the wood with a colorful latex paint and let dry.

4. Drill pilot holes to screw in the five eyelet screws (see photos for position).

5. Using a screwdriver and hammer, punch out a ¾-inch slug from opposite sides of the junction box. Position the junction box on the free end of the wood base (with one of the open holes facing the upright wooden arms) and attach with the two ¾-inch wood screws. Attach the two clamp combination connectors through the two holes in the junction box and tighten in place with pliers.

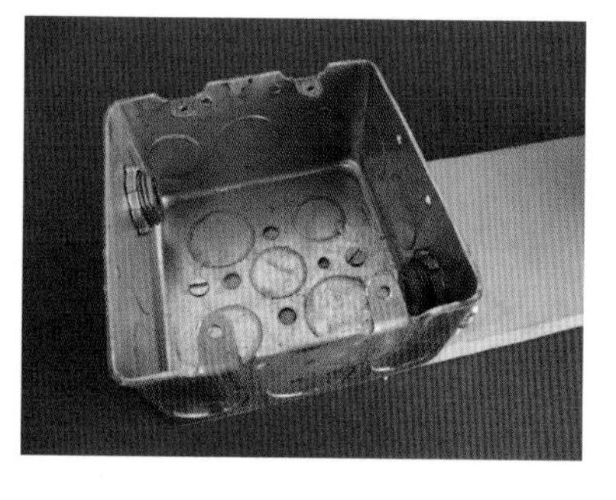

6. Insert the two 3-inch screws through the center holes in the wood arms so that the screws are facing inward. Make certain the ends of the screws cannot possibly touch each other. If you have built the device properly, the ends of the screws should be 2 inches apart. Push the screws back so they are halfway through the holes.

7. Using the wire cutters, strip 2 inches of plastic coating off one end of the 26-inch-long wire and one end of the 14-inch-long wire. Wrap each exposed wire around its own screw, just below the head. Strip ½ inch of plastic coating off the opposite ends of

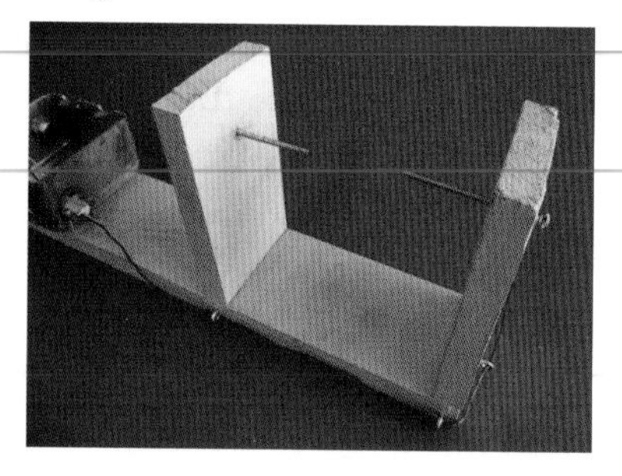

the two wires. Thread each wire through the appropriate eyeholes and through the facing clamp combination connector. Insert a small piece of folded foam or cardboard into the space above the two wires to fill the gap, and then tighten the clamp into place.

8. Attach the free end of one of the wires to the bottom screw on one of the on/off switches. Attach the free end of the remaining wire to the bottom screw on the second on/off switch.
9. Using the wire cutters, cut off the outlet at the end of the extension cord, separate the two wires for 6 inches, and strip ¾ inch of plastic coating off the end of each wire.

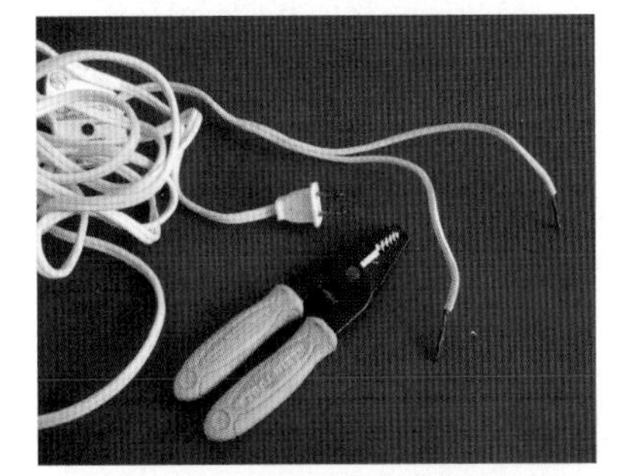

10. Thread the two wires through the second clamp combination connector to the spot where the two wires are sealed together, and then tighten the clamp into place. Attach the free end of one of the wires to the top screw on one of the on/off switches. Attach the free end of the remaining wire to the top screw on the second on/off switch.

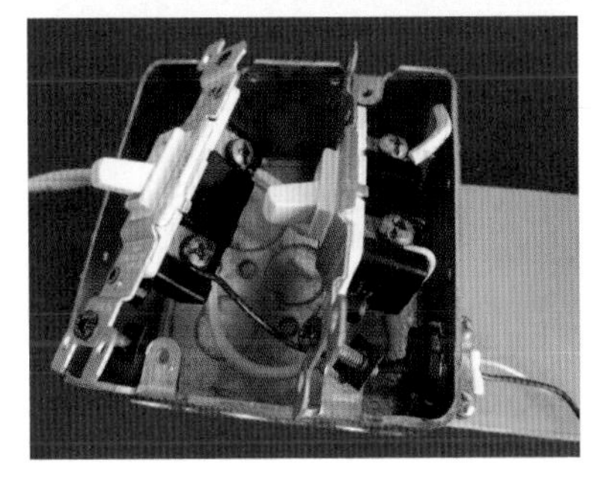

11. Screw the on/off switches to the junction box, and then screw on the plastic faceplate.
12. Before starting this experiment, make certain that the extension cord is unplugged and that both on/off switches are in the off position.
13. Place the sponge on the wood base under the two screws to absorb drips from the pickle.
14. Wearing safety goggles and rubber-soled shoes, hold the pickle horizontally between the two screws and insert them into the opposite ends of the pickle.

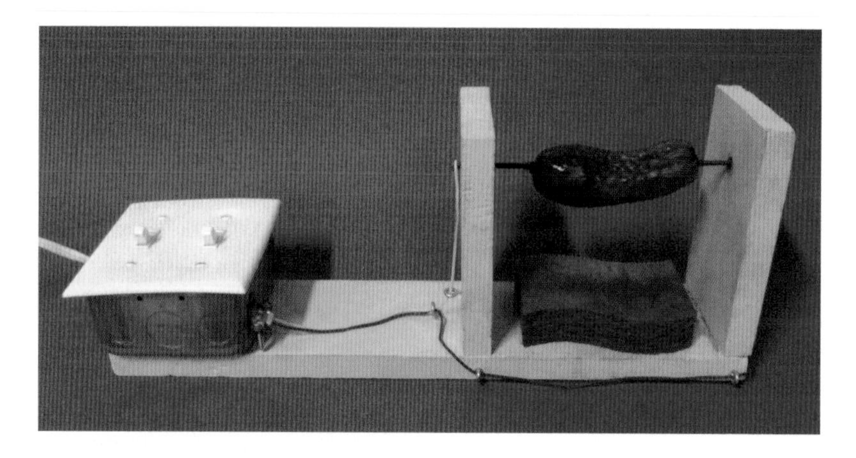

15. Make certain both on/off switches are in the off position. Plug the extension cord into an outlet. Wearing safety goggles and rubber-soled shoes, turn on the two switches. **Do not touch the pickle or either electrode when the electricity is turned on.** Otherwise, you risk giving yourself an electrical shock. The electric pickle will smell, so allow plenty of ventilation.
16. Always turn off the two switches and unplug the extension cord before removing the pickle or touching the two screws.

WHAT HAPPENS

After sputtering, hissing, and smoking briefly, the pickle glows like a bright-green electric light (usually at just one end).

WHY IT WORKS

The brine—made from vinegar (acetic acid) and salt (sodium chloride)—inside the pickle conducts electricity. The electric current flows through the pickle and heats the brine inside the pickle above the boiling point, turning the liquid to gas and causing the pickle to sizzle, steam, and give off a pungent odor. The heat eventually creates a gas-filled cavity between the electrodes inside the pickle. Sodium ions in the vapor state attach electrons from the flowing electrical current, causing an electric arc to jump the gap, much like lightning, and the pickle glows like a yellow/orange lightbulb. In other words, the electricity neutralizes the sodium ions, creating excited sodium ions that have been pushed to a higher energy level. As the sodium ions return to their ground state, they emit a frequency of light. The color of the light emitted from the pickle is the same color emitted from sodium-vapor lamps used to illuminate many parking lots.

WACKY FACTS

- A salt molecule (sodium chloride) is a sodium atom and a chlorine atom joined by an ionic bond, meaning the sodium atom gave one electron to the chlorine atom, resulting in a positively charged sodium ion and a negatively charged chloride ion.
- The scientific explanation for the glowing pickle phenomenon was first published by J. R. Appling, F. J. Yonke, R. A. Edginton, and S. L. Jacobs in an article titled "Sodium D Line Emission from Pickles" in the *Journal of Chemical Education* in 1993.
- Pickled foods are soaked in solutions that help prevent spoilage. Pickles are generally preserved in vinegar, a strong acid in which few bacteria can survive, or salt brine, which encourages fermentation—the growth of beneficial bacteria that make a food less vulnerable to harmful bacteria that cause spoilage.

Is Sodium Worth Its Salt?

- In 1807, English chemist Sir Humphry Davy first isolated pure sodium through the electrolysis of caustic soda (NaOH).
- The name *sodium* stems from the English word *soda* and the Medieval Latin word *sodanum*, meaning "headache remedy."
- Sodium's atomic symbol (Na) is derived from the Latin word *natrium*, meaning "sodium carbonate."
- Pure sodium is a highly reactive element that can ignite on contact with water.
- Used in streetlights, sodium vapor produces a brilliant yellow light.

19. Hot Ice Insanity

SAFETY FIRST

- Rubber gloves
- Safety goggles

WHAT YOU NEED

- Measuring cup
- White vinegar, 1 gallon
- Pot, 3 quarts or larger
- Measuring tablespoon
- Baking soda
- Spoon
- Hot plate or stove
- Soup bowl, microwave safe
- Aluminum foil
- Refrigerator
- Water
- Microwave oven
- 3 plastic cups
- Isopropyl alcohol, 70 percent (available at drugstores)

WHAT TO DO

1. Wearing safety goggles and rubber gloves, pour 4 cups of vinegar in the large pot.
2. Slowly add 4 tablespoons of baking soda to the vinegar, 1 tablespoon at a time, stirring well, until the bubbling and foaming reaction settles down and the liquid is clear. The bubbling is the carbon dioxide being released from the chemical reaction, leaving a solution of sodium acetate and water.

3. Place the pot over medium heat on the hot plate or stove, and let the solution boil with a rolling boil (not a rapid boil) to boil out nearly all the excess water from the mixture (roughly 1 to 2 hours).
4. When nearly all the liquid is gone and a thin, crusty film of crystals begins to form on the surface of the remaining liquid, turn off the heat and add ⅓ cup of vinegar.

5. Swirl the hot solution, and pour the yellow liquid into the soup bowl.
6. Cover the bowl with aluminum foil and let it sit undisturbed to cool slowly to room temperature.
7. Once the solution has cooled to room temperature, place the covered bowl in the refrigerator for 1 hour.
8. Use a spoon to scrape a few crystals from the pot into a plastic cup for later.

9. Crystals of sodium acetate will form in the bowl. If not, stir for a moment to trigger crystallization. Note that **sodium acetate may be harmful if swallowed or inhaled**, and the chemical may cause irritation to skin and eyes.
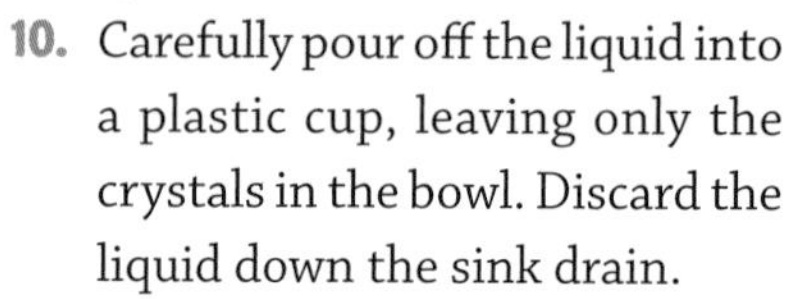
10. Carefully pour off the liquid into a plastic cup, leaving only the crystals in the bowl. Discard the liquid down the sink drain.
11. To purify the crystals, pour enough isopropyl alcohol into the bowl to cover the crystals, stir well, and carefully pour off the liquid into a plastic cup. Discard the liquid down the sink drain. Repeat this procedure two more times.
12. Add 1 tablespoon of water to the crystals, and microwave for approximately 60 seconds to melt the crystals into a liquid.
13. Pour the liquid into a plastic cup, cover with a sheet of aluminum foil (to prevent any dust or particles from contaminating the crystals), and let the solution cool slowly to room temperature.
14. Once the solution has cooled to room temperature, place the covered cup in the refrigerator for 1 hour.
15. Place a sheet of aluminum foil on a tabletop. Place the previously saved crystals on the sheet of foil.
16. Slowly pour the cup of liquid over the crystals on the sheet of foil.

WHAT HAPPENS

The liquid poured from the cup instantly turns into a stalagmite of crystals, which look like instant ice but feel astonishingly hot.

WHY IT WORKS

The reaction between baking soda (sodium bicarbonate) and vinegar (5 percent acetic acid) produces sodium acetate, water, and carbon dioxide gas. Boiling the liquid removes most of the water, creating a supersaturated solution (by dissolving more sodium acetate in the water than the water can normally hold). The seed crystals of sodium acetate initiate crystallization, and the rapid crystallization of the supersaturated solution is an exothermic process, producing heat.

WACKY FACTS

- You can reheat the crystals for reuse, dissolving them back into liquid form. Be sure to keep the solution free from dust or other particles, which can initiate premature crystallization.
- Commercial heat packs sold under the brand names ThermaCare and CardinalHealth contain a supersaturated solution of sodium thiosulphate. Squeezing and kneading the package releases a seed crystal that triggers the solution to crystallize exothermically. The heat pack can be reused by heating it to redissolve the sodium thiosulphate.
- You can also create hot ice using sodium thiosulfate. Using a pair of scissors, carefully open a heat pack containing sodium thiosulfate and pour the liquid contents into a clean bowl. Remove the packet containing the seed crystal for use as the catalyst to initiate crystallization, which produces heat. When heated, the crystals will dissolve in their own water for reuse. **Sodium thiosulfate may be harmful if swallowed or inhaled, and the chemical may cause irritation to skin, eyes, and the respiratory tract.**

20. Perplexing Crystals

SAFETY FIRST

- Safety goggles
- Rubber gloves
- Disposable dust mask

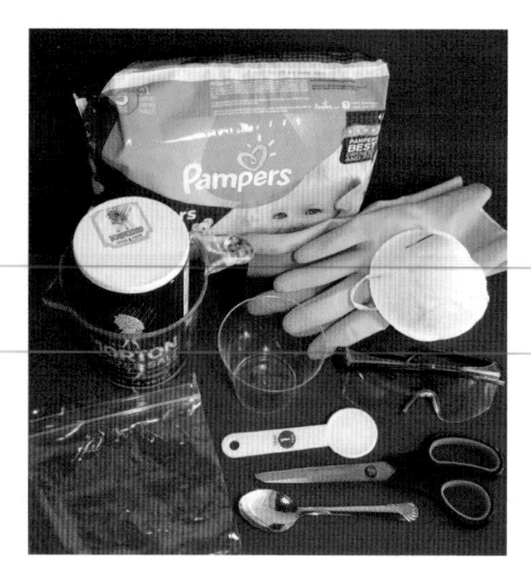

WHAT YOU NEED

- Disposable diaper
- Scissors
- Ziplock freezer bag, gallon size
- 2 clear plastic cups, 9 ounces each
- Measuring teaspoon
- Measuring cup

- Warm distilled water
- Ziplock sandwich bag
- Salt
- Spoon

WHAT TO DO

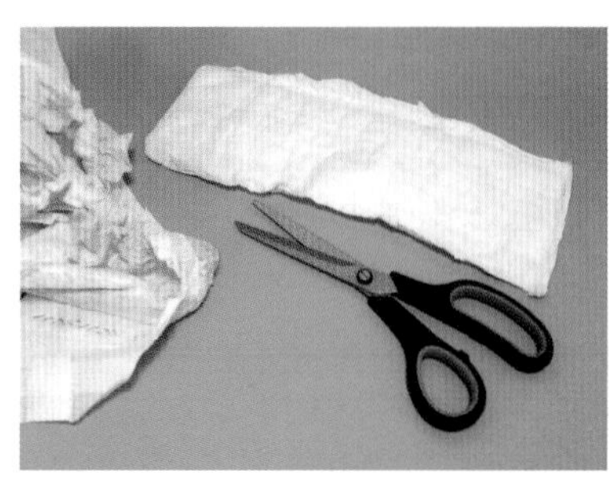

1. Wearing safety goggles, rubber gloves, and the disposable dust mask, flatten the disposable diaper and use the scissors to carefully cut off the plastic edge around the entire diaper, saving only the padded middle section. The powdered crystals (approximately the size of grains of salt) in the diaper are sodium polyacrylate, which **if inhaled, will irritate nasal membranes. Avoid eye contact, which will cause dryness and irritation.**
2. Place the padded middle section of the diaper inside the ziplock freezer bag.
3. Seal the bag shut, and from outside the plastic bag, pull open and separate the cotton, paper, and plastic layers of the diaper.
4. Shake the sealed bag for 1 or 2 minutes (or rub the grainy sheet inside the bag), until small white granules cease falling from inside the diaper to the bottom of the bag.

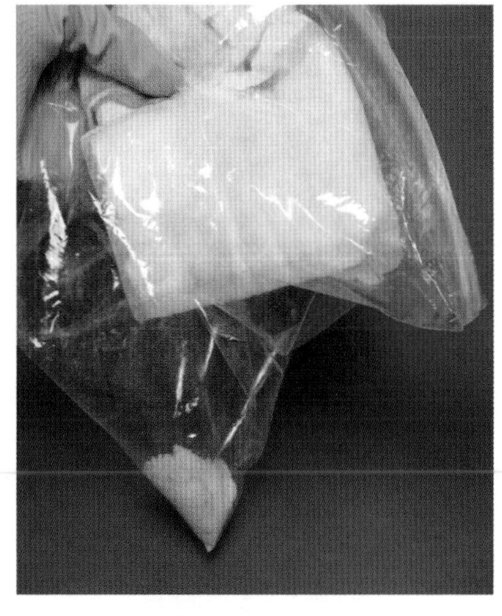

5. Without opening the bag, use your fingers on the outside of the plastic to move the pieces of cotton and plastic toward the top of the bag, and holding them in place, shake the bag gently to free any granules trapped in the cotton.
6. When approximately 1 teaspoon of granules sits isolated in a tilted corner of the bag, carefully open the bag and discard any large pieces of cotton and plastic.
7. Carefully pour the granules from the bag into a plastic cup.

8. Seal and discard the plastic bag, and wash and dry your hands thoroughly.
9. Place 1 teaspoon of the granules in the second plastic cup. Add 1 cup of warm distilled water and observe. Let the cup sit undisturbed for 2 to 8 hours for the crystals to reach their maximum size.
10. Save the remaining dry crystals in a ziplock sandwich bag.
11. Add 1 teaspoon of salt to the cup of crystals. Stir well, let sit for 1 hour, and observe.

WHAT HAPPENS

When placed in water, the granules of sodium polyacrylate absorb water and swell to several times their original size. When salt is added to the cup of crystals, the crystals dehydrate into flaky granules and sink to the bottom of the cup of salt water.

WHY IT WORKS

The powdered granules in disposable diapers are sodium polyacrylate—a nontoxic, superabsorbent, hydrophilic ("water loving") polymer that absorbs up to 800 times its weight in water. If placed in water, sodium polyacrylate absorbs water through osmosis and swells to several times its original size. The polymer cannot dissolve due to its three-dimensional polymeric network structure. The amount of water this polymer absorbs depends on the salt content of the water. When a teaspoon of salt (sodium chloride) is added to the water, the distilled water in the crystals, being a less concentrated solution than the salt water in the cup, passes through the semipermeable crystal membrane into the salt water. In other words, the salt dehydrates the crystals.

WACKY FACTS

- Sodium polyacrylate displays liquid-like properties because the polymer is composed almost entirely of water. The polymer also exhibits solid-like properties due to the network formed by the cross-linking reaction.
- Superabsorbent polymers are currently being used in such industries as pharmaceuticals, food packaging, paper production, horticultural, and oil drilling, for products including gelatin, disposable diapers, contact lenses, and gravy.
- Originally developed by the US Department of Agriculture from hydrolyzed starch and polyacrylonitrile, the sodium polyacrylate found in today's disposable diapers is made from sodium salts cross-linked with polyacrylic acid.
- Sodium polyacrylate will absorb approximately 800 times its own weight in distilled water, approximately 300 times its own weight in tap water, and about 60 times its weight in a 0.9 percent salt solution (similar to the salt concentration of urine).
- Sodium polyacrylate can be dehydrated and rehydrated repeatedly. To bring the crystals back to their original size, spread the expanded crystals on a flat surface and let dry.
- Superabsorbent crystals are sold at garden centers under several brand names, including HydroSource or SoilMoist.
- Sodium polyacrylate crystals are used by florists in planters, by firefighters in combating fires, and in disposable diapers.

21. Baffling Money Burn

SAFETY FIRST

This experiment should be performed outside.

- Metal tongs
- Safety goggles
- Fire extinguisher

WHAT YOU NEED

- Measuring cup
- Isopropyl alcohol, 91 percent (available at drugstores)
- Water
- Two soup bowls
- ½ teaspoon measuring spoon
- Salt
- Spoon
- $1 bill (or if you're feeling more adventurous, a larger denomination)
- Butane barbecue lighter

WHAT TO DO

1. Mix 2 ounces of isopropyl alcohol and 2 ounces of water in a soup bowl.
2. Add ½ teaspoon of salt to the solution and stir well to dissolve the salt.
3. Immerse the $1 bill in the solution, making sure the currency is saturated.
4. Using the tongs, remove the bill from the bowl. Squeeze out any

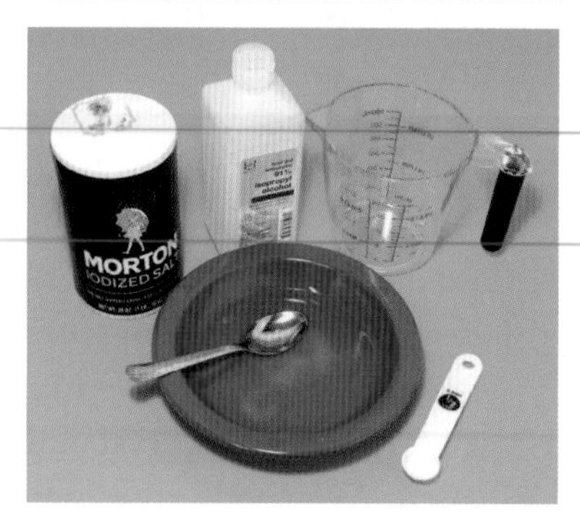

excess liquid so the solution no longer drips from the bill.

5. Place the bowl containing the highly flammable alcohol-water solution at a safe distance from where you plan to ignite the bill.
6. Place a second bowl containing only water near where you plan to ignite the bill.
7. Wearing safety goggles and with a fire extinguisher on hand, use the metal tongs to hold one end of the bill, and using the butane barbecue lighter, ignite the other end of the bill.

8. If the bill actually begins to burn, submerge it in the bowl of water.

WHAT HAPPENS

A bright blue-and-yellow flame appears to burn the bill. When the flame burns off all the alcohol, the fire extinguishes itself and the bill remains unscathed and cool to the touch.

WHY IT WORKS

The water (a compound of hydrogen and oxygen) in the solution insulates the bill and prevents the paper money from burning along with the alcohol. Alcohol burns with a nearly invisible blue flame, but the salt makes the flame more visible. The water absorbs most of the heat energy from the flame, preventing the paper from catching fire. The heat from the burning alcohol boils and then vaporizes the water—but not before the alcohol burns off.

WACKY FACTS

- You can also conduct this experiment using 95 percent ethyl alcohol (mix 2 ounces with 2 ounces of water) or 70 percent isopropyl alcohol (mix 7 tablespoons with 3 tablespoons of water).

- If saturated with a pure alcohol solution, the bill will burn.
- Reducing the amount of water in the solution makes the paper money more likely to char or catch fire.

IT'S ELEMENTARY

The Cobalt Malt

German miners, cursing the dark-blue ore that hindered them from extracting silver, named the troublesome ore *kobold*, meaning "goblin." In the 15th century, German glassmakers used the ore to give glass and pottery a beautiful blue color. Launders used kobold ore as bluing to make clothes appear whiter.

In 1730, Swedish chemist Georg Brandt, head of the Bureau of Mines in Stockholm, discovered that the ore contained a previously unknown metal element, which he named kobalt. In English, the ore became known as cobalt. For many years, chemists disputed Brandt's discovery, insisting that the new element—a lustrous, magnetic, silvery-blue metal—was nothing more than a compound of iron and arsenic. Finally, around 1780, Swedish chemist Torbern Bergman confirmed Brandt's discovery that cobalt was an element in its own right.

Cobalt oxides are used as pigments to produce brilliant blue colors in paints, enamel, pottery, and glass. The tomb of King Tut, who ruled Egypt from 1361 to 1352 BCE, contained a small glass object colored deep blue with cobalt. Cobalt is alloyed with aluminum and nickel to make powerful magnets. Other cobalt alloys that withstand high temperatures are used in jet turbines and gas turbine generators. Radioactive cobalt-60 is used to treat cancer and to irradiate food to kill bacteria.

MAGNESIUM

22. Mystifying Milk of Magnesia

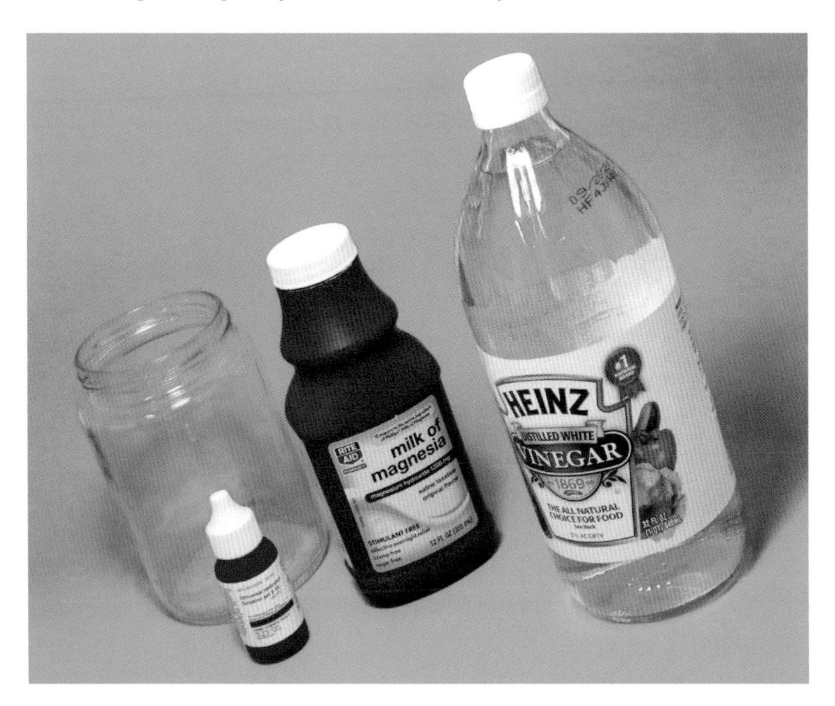

SAFETY FIRST

- Safety goggles
- Rubber gloves

WHAT YOU NEED

- Measuring cup
- Milk of magnesia, containing 1,200 milligrams magnesium hydroxide (available at drugstores)

- 24-ounce glass jar, clean and empty
- Distilled water
- Measuring teaspoon
- Universal indicator (available at chemical supply stores or www.homesciencetools.com)
- Spoon
- Glass or metal cake pan, 8-inches square
- Ice
- Water
- Vinegar, 5 percent acetic acid

WHAT TO DO

1. Wearing safety goggles and rubber gloves, shake the bottle of milk of magnesia well, pour 2 ounces of milk of magnesia into the glass jar, and add 3 ounces of distilled water.
2. Add 1 teaspoon of universal indicator. Mix well with a spoon.
3. Place the jar in the cake pan, and fill the cake pan with ice and water. Wait 5 minutes for the ice water to cool the solution in the jar.

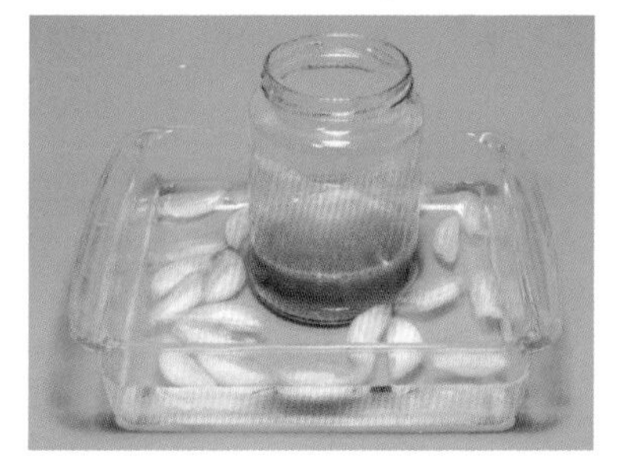

4. While swirling the solution in the jar, add 1 ounce of vinegar and observe.
5. When the solution turns bluish purple, add another ounce of vinegar to the solution, swirling as you do, until the solution stops changing colors.
6. Repeat step 5 until the end result is a clear red solution.
7. When you finish observing, discard the solution by flushing it down a sink drain with ample water.

WHAT HAPPENS

When the universal indicator is mixed with the milk of magnesia, the solution turns bluish purple. When vinegar is added to the solution, the mixture immediately turns red and then a series of colors—orange to yellow to green to a bluish-purple color. Eventually, after adding vinegar repeatedly, the solution turns red and clear.

WHY IT WORKS

Universal indicator turns red when added to an acidic solution and dark purple when added to an alkaline (or basic) solution. When the indicator is added to the milk of magnesia suspended in water, the solution turns light purple, indicating its basic nature at pH 9. The ice cools the solution, slowing the reaction to make it easier to observe.

When vinegar is added to the solution, the acetic acid in the vinegar neutralizes the small amount of dissolved hydroxide ion from the milk of magnesia, turning the solution acidic and momentarily red, indicating the acidic nature of the vinegar.

As the magnesium hydroxide from the suspension gradually dissolves in the solution, neutralizing the acetic acid, the liquid changes through the entire universal indicator color range, ultimately turning the solution basic (purple).

As you add more vinegar to the solution, the colors change more slowly, and as the solution becomes more acidic, the color stops at blue (pH 8), green (pH 7), yellow (pH 6), orange (pH 5, pH 4, and pH 3), and finally red (pH 2).

WACKY FACTS

- Milk of magnesia (a suspension of magnesium hydroxide in water) reacts with acid to form water and magnesium ions. Milk of magnesia neutralizes stomach acid by this same method.
- In 1872, British pharmacist Charles Henry Phillips, living in New York, invented Phillips' Milk of Magnesia by mixing magnesium hydroxide with water. Phillips trademarked the name Milk of Magnesia in 1880 and manufactured the antacid laxative formula at a plant in Glenbrook, Connecticut.
- You can give yourself a facial with milk of magnesia. Use a cotton ball to apply milk of magnesia as a facial mask, let dry for 30 minutes, and then rinse off with warm water followed by cool water. The milk of magnesia absorbs the oils from your skin while cooling the skin at the same time, leaving your face minty fresh.

Magnesium: The Bee's Kneesium

- Magnesium is the lightest metal that can be used in construction.
- English chemist Sir Humphry Davy discovered magnesium in 1908.
- Magnesium is named after the district of Magnesia in Greece, where the compound magnesium carbonate was first discovered.
- Pure magnesium does not occur in nature. Instead, various minerals contain magnesium compounds, most notably magnesium chloride and magnesium sulfate.
- Magnesium plays a vital role in plant photosynthesis and the duplication of DNA and RNA, and activates many of the enzymes that speed up chemical reactions in the human body.

23. Crazy Colored Crystals

WHAT YOU NEED

- Measuring cup
- Water
- Small microwave-safe glass mixing bowl
- Microwave oven
- Epsom salt
- Spoon
- Food coloring

WHAT TO DO

1. Place ½ cup of water into the small microwave-safe glass mixing bowl. Heat the water in the microwave oven for 45 seconds.
2. Mix ½ cup of Epsom salt with the ½ cup of hot water in the mixing bowl, and stir for at least 2 minutes. (Some undissolved crystals of Epsom salt will likely remain at the bottom of the bowl.)

3. Add 3 drops of whatever color of food coloring you wish your crystals to be, and mix well.
4. Place the bowl in the refrigerator and let it sit undisturbed overnight (or longer if desired).
5. Gently remove the bowl from the refrigerator and carefully drain any remaining liquid down the sink.

WHAT HAPPENS

You've created a bowl full of needlelike crystals.

WHY IT WORKS

The temperature of the water determines how much Epsom salt (magnesium sulfate) dissolves in the water. The hotter the water, the more magnesium sulfate it can hold. Dissolving Epsom salt in hot water creates a saturated solution, meaning no more magnesium sulfate can dissolve in the water. Cooling the solution in the refrigerator triggers rapid crystallization because the cooler solution cannot hold the excess magnesium sulfate. As the solution cools (becoming denser), the magnesium sulfate molecules join together in hundreds of slender crystals.

WACKY FACTS

- Left undisturbed, the crystals created in this experiment can last for several months or longer.
- Epsom salt is named after the springs in Epsom, England, where it was first mined, and for the fact that the crystals look like salt crystals.
- According to the 1662 book *Worthies* by Thomas Fuller, in 1618 a farmer in Epsom, England, discovered that during a drought his cows refused to drink the water from a mineral spring. Discovering that the water tasted bitter, the farmer soon noticed that it helped heal scratches and rashes on his skin. The spring developed into Epsom Wells, a spa frequented by people from London, including such luminaries as John Aubrey, Samuel Pepys, and Prince George of Denmark. In 1675, the mineral contents were extracted from the spring. The resulting salt was magnesium sulfate, which, when combined with water, produced a substance with laxative properties.

- Magnesium, the second most abundant element in human cells, helps to regulate the activity of more than 325 enzymes and performs a vital role in coordinating muscle control, electrical impulses, energy production, and the elimination of harmful toxins.
- According to the National Academy of Sciences, most Americans are magnesium deficient, and this deficiency causes heart disease, stroke, osteoporosis, arthritis and joint pain, digestive maladies and stress-related illnesses, chronic fatigue, and many other ailments.
- The human body can absorb Epsom salt, dissolved in a bath, through the skin, replenishing the body's levels of magnesium. Soaking in an Epsom salt bath also relieves stress by raising the body's level of serotonin, lowering the effects of adrenaline, helping to regulate the electrical impulses in the nerves, and lowering blood pressure.

ALUMINUM

26.9815386 2 8 3
Al
13

24. Incredible Shrinking Potato Chip Bag

SAFETY FIRST

- Oven mitt

WHAT YOU NEED

- Empty potato chip bag
- Wax paper
- Microwave oven
- Pie pan
- Scissors
- Hole puncher
- Key ring

WHAT TO DO

1. Flatten the empty potato chip bag.
2. Fold a sheet of wax paper in half, and place the potato chip bag between the folded sheet.
3. Place the wax-paper folder, containing the empty potato chip bag, in the microwave oven.
4. Heat for 5 seconds.
5. Wearing an oven mitt, immediately and carefully remove the wax-paper folder containing the empty potato chip bag from the microwave oven (without touching the hot plastic bag), and immediately use the bottom of the pie pan to press the hot plastic flat (through the wax paper).

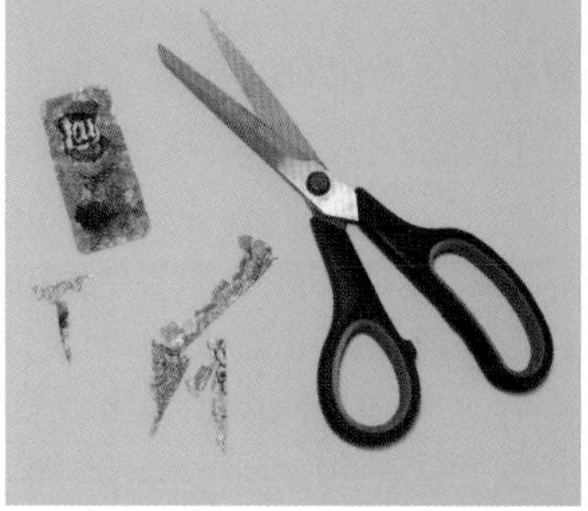

6. Let cool for 1 minute.
7. Remove the plastic bag from the wax paper.
8. Using the scissors, carefully trim the edges of the plastic bag and round the corners.
9. Use the hole puncher to make a hole in the plastic, and attach a key ring to the hole.

WHAT HAPPENS

Sparks shoot from the potato chip bag. You'll hear a metallic crinkling noise, and when you remove the bag from the microwave oven, it will be a miniaturized version of itself—approximately one-eighth its original size.

WHY IT WORKS

The chip bag is composed of a thin sheet of aluminum foil sandwiched between two thin films of plastic. The aluminum prevents air from entering the package and turning the chips rancid. The plastic film is made from a polymer (a long string of molecules) that has been stretched out, like an elongated Slinky. The heat from the microwave oven causes the molecules in the polymer to vibrate faster, reverting to their natural state (or springing back together like the coils of a Slinky), which shrinks the package.

WACKY FACTS

- Potato chip bags are filled with nitrogen, which preserves the freshness of the chips, prevents combustion, and creates sufficient cushioning during shipping so the chips don't get crushed.
- The National Aeronautics and Space Administration (NASA) reports that bags of potato chips taken aboard supermodified jets respond to the sudden change in air pressure soon after takeoff by exploding.
- If you light a candle, cut off the top corner of an unopened bag of potato chips, and gently squeeze the gas from the potato chip bag toward the base of the candle, the candle flame goes out. The nitrogen from the bag floods the area around the candle, pushing away the oxygen needed for the candle to burn.

25. Kooky Can Crusher

SAFETY FIRST

- Metal tongs

WHAT YOU NEED

- Pot, 3 quarts or larger
- Water
- Measuring tablespoon
- Aluminum soda can, clean and empty
- Stove

WHAT TO DO

1. Fill the pot with cold water.
2. Pour 1 tablespoon of cold water into the aluminum soda can.
3. Set the can on the burner of a stove, and heat the can to boil the water inside the can. When the water comes to a boil, a cloud of condensed vapor will suddenly escape from the opening in the can. Let the water boil for approximately 30 seconds. Do not heat the can over high heat or when it is empty. If you do, the ink printed on the can may burn or the aluminum may melt.
4. Using the metal tongs, carefully grasp the can and quickly turn it upside down and submerge it in the water in the pot.

WHAT HAPPENS

The can will instantly collapse as if magically crushed.

WHY IT WORKS

When the water boils inside the can, the steam forces the air out of it. When the aluminum can is placed in the cold water, the steam condenses back into liquid water, creating a partial vacuum inside the can. The air pressure outside the can instantly crushes it.

WACKY FACTS

- The air pressure on any object at sea level is 14.7 pounds per square inch.
- Normally the air pressure inside an open can equals the air pressure outside the can.
- Gravity causes the weight of the air pressing down on the Earth, the ocean, and on the air below to create air pressure.
- The amount of air pressure in any one point depends on the amount of air above that point. Air pressure (and air density) decreases with increased altitude.
- Changes in air pressure can affect your body, most noticeably when flying aboard an airplane, riding a rapid elevator, driving up or down mountain roads, or swimming several feet underwater.

Aluminum: Foiled Again

- Danish chemist Hans Christian Orsted extracted the first aluminum from the mineral alum in 1825.
- The name of the element aluminum is derived from the mineral alum, named from the Latin word *alumen*, meaning "bitter salt."
- In English-speaking countries other than the United States, aluminum is called aluminium.
- Aluminum does not rust.
- Aluminum is the third most used metal in the world, after iron and steel.

SILICON

28.0855	2 8 4
Si	
14	

26. Magical Mystery Sand

SAFETY FIRST

This experiment should be performed outside.

- Safety goggles
- Rubber gloves
- Disposable dust mask

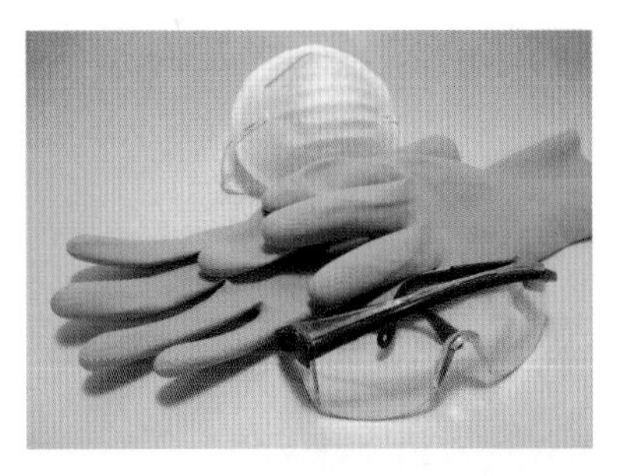

WHAT YOU NEED

- Cookie sheet
- Wax paper or baking paper
- 1½ cups of play sand
- Scotchgard Fabric Protector (available at office supply and craft stores)
- Fork
- Clear plastic cup

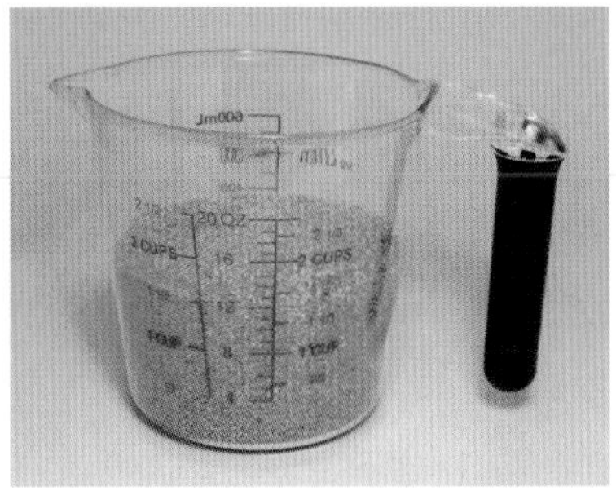

- Glass mixing bowl
- Water
- Airtight plastic container

WHAT TO DO

1. Line a cookie sheet with a sheet of baking paper or wax paper.
2. Spread out a thin layer of sand on the baking paper or wax paper.
3. Working outdoors in a well-ventilated area and wearing safety goggles, rubber gloves, and a disposable dust mask, spray a heavy coat of Scotchgard Fabric Protector on the granules of sand. Gently shake the pan several times to move the sand around while you spray it.
4. Let the sand dry for 1 hour or longer.
5. Use a fork to rake the sand and repeat steps 3 and 4 above.
6. When the sand dries completely, pour it from the sheet of wax paper into a clear plastic cup.

7. Fill the glass mixing bowl ¾ full of water.
8. Slowly pour the prepared sand from the cup into the bowl of water, observing how the sand reacts with the water.
9. When you're finished, reach into the bowl and scoop out the sand.
10. Store the sand in a sealed container for future use.

WHAT HAPPENS

When poured into the bowl of water, the sand creates underwater sculptures. When scooped out of the water, the sand remains completely dry.

WHY IT WORKS

When sprayed on sand, Scotchgard Fabric Protector, a waterproofing agent used on fabrics and other materials to create a coating that repels water, makes the sand hydrophobic ("afraid of water"). In other words, the sand (composed of silicon dioxide), coated with a water-repelling substance, can no longer absorb or mix with water, keeping it dry.

WACKY FACTS

- In 1944, 3M bought the rights to a process for producing fluorochemical compounds. Researchers at 3M could not find any practical uses for the process or its reactive, fluorine-containing by-products—until a laboratory assistant accidentally spilled a sample of the substance on her tennis shoes. The assistant could not wash the stuff off with water or hydrocarbon solvents, and that spot on her tennis shoe also resisted soiling. 3M chemists Patsy Sherman and Sam Smith realized this substance might be used to make textiles resist water and oil stains and went to work to enhance the compound's ability to repel liquids, giving birth to Scotchgard.
- Hydrophobic sand can be used in potted plants to allow the roots to breathe, even when the plant has been overwatered. A layer of hydrophobic sand placed in the bottom of a potted plant pot prevents water from passing through it, yet allows air to pass through the sand grains.
- Landscapers can use hydrophobic sand to create a natural-looking stream, with water flowing over sand that appears normal.
- Agriculture farmers can place a layer of hydrophobic sand beneath the soil to conserve water by creating a man-made water table. The hydrophobic sand stops the water from seeping into the groundwater. Instead, the water lies trapped above the hydrophobic sand layer, creating an artificial water table that provides sustenance for plant roots.
- Hydrophobic sand was invented to clean up oil spills in water. When poured on an oil spill, the coated sand bonds to the oil and sinks to the bottom, where it can be dredged and treated. Unfortunately, the cost of using hydrophobic sand in this manner is prohibitively expensive.
- In cold climates, hydrophobic sand has been used to cover electrical junction boxes buried in the ground. To make repairs, workers can dig through the hydrophobic sand and unearth the junction boxes, even when the ground is frozen.

Hide Your Head in the Silicon

- Swedish chemist Jöns Jacob Berzelius first isolated silicon in 1824.
- English chemist Sir Humphry Davy proposed the name *silicium* for the element, from the Latin word *silicis*, meaning "fling," and the suffix *-ium*, denoting a metal. However, in 1817, Scottish chemist Thomas Thomson named the element silicon, retaining part of Davy's proposed name but changing the suffix to *-on*, correctly identifying the element as a nonmetal (like boron and carbon).
- Silica, an oxide of silicon, is the most common ingredient of sand.
- According to the Royal Society of Chemistry, silicon is the seventh most abundant element in the universe and the second-most abundant element on the planet, after oxygen.
- Silicon is used to make transistors, solar cells, and computer chips.
- Silicon Valley is nicknamed for the silicon used in computer chips. The sobriquet first appeared in 1971 in the newspaper *Electronic News*.
- Silicon and silicone are two different things. Silicone, the polymer used in breast implants, is a synthetic substance made from silicon, oxygen, carbon, and hydrogen.

PHOSPHOROUS

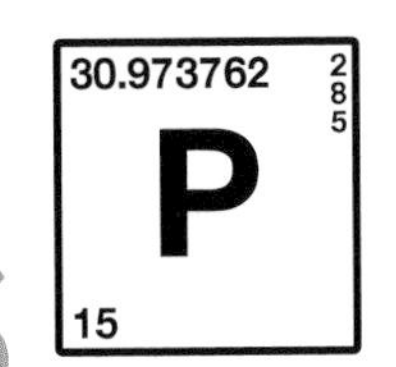

27. Radical Rust Remover

WHAT YOU NEED

- Drinking glass
- Coca-Cola
- Rusty iron nails

WHAT TO DO

1. Fill a drinking glass with Coca-Cola.
2. Place the rusty nails in the liquid.
3. Let sit undisturbed for 1 hour.
4. Carefully cover the mouth of the glass with your hand and pour the liquid through your fingers and down a sink drain, catching the nails from the glass.
5. Discard the used cola down the sink drain.
6. Observe the nails.

WHAT HAPPENS

When the nails are placed in the Coca-Cola, small bubbles form on the nails and immediately rise to the surface. Ultimately, the cola removes all the rust from the nails.

WHY IT WORKS

The iron nails rust when the iron oxidizes. In other words, the iron combines with the oxygen in the air to form iron oxide (rust). The phosphoric acid (a colorless, odorless, inorganic acid containing phosphorus) in the cola dissolves the iron oxide from the nails. The phosphoric acid simultaneously and slowly etches the iron—causing little damage to the nails—and also leaves a thin protective coating of iron phosphate on the nails.

WACKY FACTS

- Phosphoric acid is a clear, colorless, odorless liquid with the consistency of syrup.
- The phosphoric acid makes Coca-Cola more acidic than lemon juice or vinegar.
- Phosphoric acid is commonly used in rust-removal products, such as Naval Jelly.
- Phosphoric acid contained in cola soft drinks has been linked to lower bone mineral density in women, according to a 2006 study published in the *American Journal of Clinical Nutrition*.

Phosphorus Ain't Preposterous

- In 1669, German alchemist Hennig Brand, working in Hamburg, attempted to make gold by letting a jar of urine stand for several days to putrefy. He boiled the rotting liquid down to a paste, heated the resulting paste, and drew the vapors into water. Instead of condensing to gold, as Brand anticipated, the drawn vapors produced a white, waxy substance that glowed in the dark. He named his accidental discovery phosphorus (Greek for "light bearer"), marking the first time anyone had discovered an element unknown to ancient peoples.
- In 1830, French chemist Charles Sauria developed phosphorus matches, inadvertently giving rise to "phossy jaw," a fatal disease technically known as phosphorus necrosis and primarily affecting factory workers. Highly poisonous, phosphorus essentially causes bones to disintegrate and deform. The phosphorus scraped from the heads of a pack of matches could also be used to commit suicide or murder.
- Phosphates, used in detergents to soften water, contain the element phosphorus, one of many nutrients essential to water plants and algae. Since phosphorous also contributes to accelerated eutrophication—the excessively rapid growth of aquatic plant life in bodies of water—many state governments have banned the sale of detergents with phosphates.

28. Black Light Jell-O

SAFETY FIRST

The tonic water in this experiment will be very hot.

WHAT YOU NEED

- Tonic water, 1 liter (available at grocery stores)
- Microwave-safe measuring cup
- Microwave oven
- One packet of Jell-O, 3 ounces (any flavor)
- Glass mixing bowl
- Mixing spoon
- 4 clear whiskey/juice drinking glasses or glass dessert bowls
- Black light

WHAT TO DO

1. Pour 1 cup of cold tonic water into the microwave-safe measuring cup, and heat for 45 seconds or until boiling in the microwave oven.
2. Pour the packet of Jell-O into the glass mixing bowl.
3. Carefully pour the boiling tonic water into the bowl, and using a mixing spoon, stir for 2 minutes or until the gelatin mix is completely dissolved and the fizzing stops.
4. Add 1 cup of cold tonic water, and stir until the fizzing stops.
5. Pour equal amounts (½ cup) of the Jell-O solution into four separate drinking glasses or glass dessert bowls.
6. Refrigerate for 4 hours or until firm.
7. Place the bowls of Jell-O on a countertop in a dark room, turn on the black light, and observe. The Jell-O is safe to eat, but you might not want to if you don't like the taste of tonic water.

WHAT HAPPENS

The Jell-O glows under the black light.

WHY IT WORKS

Phosphorescent substances—like phosphorus—appear ordinary in visible light. Ultraviolet light, however, having a higher frequency and lower wavelength than visible light, excites the atoms of phosphorescent substances, making them appear to glow (or phosphoresce). The tonic water added to the gelatin mix contains quinine—a phosphorescent chemical that is activated by the ultraviolet rays produced by a black light. The quinine converts the ultraviolet light into visible light.

WACKY FACTS

- Ultraviolet rays (also known as black light) are a form of light invisible to the human eye and lie beyond the violet end of the visible spectrum.
- Highlighter pens possess fluorescent properties that absorb the energy from ultraviolet light and reemit it at lower energies, producing light frequencies in the visible spectrum.
- Both human blood and urine are fluorescent.
- Ultraviolet rays can cause sunburn and penetrate clouds, which is why a person can get sunburned on an overcast day.

IT'S ELEMENTARY
Uranium on the Cranium

In 1789 German chemist Martin Heinrich Klaproth dissolved pitchblende, a dark bluish-black mineral found in silver mines, in nitric acid and precipitated a yellow compound. He identified the compound as the oxide of a new element, which he named uranium in honor of the planet Uranus, which had been discovered eight years earlier.

More than 50 years later in 1841, French chemist Eugène Péligot isolated uranium by heating uranium tetrachloride with potassium. No one had any idea that uranium was radioactive until 1896, when French physicist Antoine Henri Becquerel, a professor at the École Polytechnique in Paris, decided to investigate whether substances made phosphorescent by sunlight might then emit a penetrating radiation similar to X-rays. Having inherited a supply of uranium salts, which phosphoresce when exposed to light, from his physicist father, Becquerel wrapped a photographic plate in opaque black paper (to shield the plate from any light), set the wrapped plate in the sunlight, and placed a uranium crystal on top of it. When he developed the photographic plate, an image of the uranium crystal appeared.

Convinced that he had proven that sunlight activated the phosphorescence of the uranium crystal (and unable to continue his experiments for several days due to overcast skies in Paris), Becquerel put the uranium crystal in a drawer on top of a photographic plate wrapped in opaque paper. Several days later, he developed the photographic plate, expecting to find a faint image of the crystal due to the waning phosphorescence of the uranium crystal. Instead, he found an image of the crystal as strong as the one he created by sitting the crystal and the wrapped plate in the sunlight. Becquerel concluded that his theory had been wrong. The uranium crystal, without the aid of sunlight, had exposed the photograph plate. He had accidentally discovered radioactivity.

Today uranium is used as nuclear fuel to generate electricity in nuclear power stations, to power nuclear submarines and spacecrafts, and in nuclear weapons.

SULFUR

32.065	2 8 6
S	
16	

29. Ghastly Green Eggs

WHAT YOU NEED

- 2 raw eggs
- Pot, 3 quarts or larger
- Water
- Hot plate or stove
- Timer
- Drinking glass
- Spoon
- Ice
- Refrigerator

WHAT TO DO

1. Place the raw eggs in a pot of water, and place the pot on a hot plate or stove burner set on high.
2. When the water begins to boil, lower the heat and simmer the eggs for 15 minutes.
3. Remove the pot from the heat, and turn off the stove.

4. Fill the drinking glass halfway with cold water.
5. Using a spoon, carefully remove one egg from the pot and place it in the glass of cold water. If necessary, add a few ice cubes to the glass to keep it cool.
6. Leave the remaining egg in the pot of hot water.
7. Let the eggs sit in the water for 30 minutes.
8. Remove both eggs from the water, and place them in the refrigerator overnight.
9. Peel the shells from the eggs, and remove the yolks. Examine and smell both yolks.

WHAT HAPPENS

The yolk of the egg that soaked in the hot water has a grayish-green coating roughly ⅛ inch thick, and the yolk of the egg that soaked in the cold water appears bright yellow.

WHY IT WORKS

The albumen (egg white) contains a protein composed of sulfur and other ingredients. When you boil an egg, the sulfur is released from the protein in the form of hydrogen sulfide gas. As you continue boiling the egg, the egg begins to heat up faster on the outside than the inside, causing the hydrogen sulfide gas to move toward the cool center of the egg. When the hydrogen sulfide gas reaches the yolk, which contains iron, it begins reacting with the iron to form iron II sulfide, which appears green. Soaking the egg in cold water makes the outside of the egg cooler than the inside, causing the hydrogen sulfide gas to migrate out through the walls of the shell and into the cool water, where it cannot react with the iron in the yolk and form iron II sulfide. This migration occurs because gas is more soluble in a cold solution than a hot one.

WACKY FACTS

- The green coating (iron II sulfide) on the yolk of an egg is edible and harmless, but most people find a bright-yellow yolk more appealing.

- Each egg yolk contains approximately 25 milligrams of sulfur, and the albumen (egg white) contains 50 milligrams, according to *Chemical & Engineering News*.
- A single eggshell is perforated by an average of 9,000 pores. The eggshell is a permeable membrane that allows air and moisture to pass through.
- Well water that stinks of rotten eggs contains high amounts of sulfur.
- "I'm frightened of eggs," horror-film director Alfred Hitchcock told Italian journalist Oriana Fallaci in 1963. "Worse than frightened—they revolt me. That white round thing without any holes—have you ever seen anything more revolting than an egg yolk breaking and spilling its yellow liquid?"
- The albumen (egg white) consists of 90 percent water, seven major proteins, and no fat.
- A rotten egg is formed when bacteria penetrate the shell and produce foul-smelling hydrogen sulfide.
- Acid rain contains diluted sulfuric acid, which is formed by the atmospheric oxidation of sulfur dioxide in the presence of water.
- The upper atmosphere of the planet Venus is primarily composed of tiny droplets of sulfuric acid, produced by the sun's photochemical action on carbon dioxide, sulfur dioxide, and water vapor.
- Sulfuric acid is used to produce fertilizers, rayon, car batteries, synthetic detergents, dyes and pigments, explosives, and drugs.

Raising a Stink over Sulfur

- Sulfur is named for the Latin word *sulphur*, meaning "to burn."
- Although sulfur has been known since ancient times, French chemists Louis-Josef Gay-Lussac and Louis-Jacques Thénard first isolated the element in 1809.
- Sulfur is mentioned 15 times in the Bible, which associates the element with fire and brimstone. "On the wicked he will rain fiery coals and burning sulfur," says Psalm 11:6. "A scorching wind will be their lot."
- Pure sulfur has no odor. The foul stench associated with sulfur actually comes from sulfur compounds, such as hydrogen sulfide.

CHLORINE

35.453 2 8 7
Cl
17

30. Wacky Woolly Wool

SAFETY FIRST

- Safety goggles
- Rubber gloves
- Disposable dust mask

WHAT YOU NEED

- Ruler
- Scissors
- Yarn, 100 percent wool (available at fabric stores)
- 24-ounce glass jar, clean and empty
- Chlorine bleach

WHAT TO DO

1. Measure and cut a 72-inch (4-foot) strand of wool yarn.
2. Place the strand of wool yarn inside the jar.
3. In a well-ventilated area and wearing safety goggles, rubber gloves, and a disposable dust mask, carefully pour ½ cup of bleach into the jar. **Chlorine bleach may cause irritation or damage to the eyes. It is harmful if swallowed. Wash your hands after contact. Avoid breathing the vapors.**
4. Let the jar stand uncovered and undisturbed for 5 to 10 minutes.

5. From outside the jar, examine the piece of wool yarn. Gently swirl the jar, which will be warm to the touch. (Do not reach inside the jar or touch the yarn.)
6. Carefully discard by pouring the solution down a sink drain, rinsing thoroughly with water.

WHAT HAPPENS

The bleach dissolves the wool yarn into a layer of white goo that floats on the surface of the liquid.

WHY IT WORKS

Chlorine bleach is a basic chemical with a pH of approximately 10. Wool is an acidic fiber. When wool is soaked in bleach, the base and acid attempt to neutralize each other, and during the exothermic chemical reaction, the bleach dissolves the wool into a white substance and generates heat.

WACKY FACTS

- Manufacturers make household chlorine bleach by mixing chlorine gas with sodium hydroxide, creating sodium hypochlorite (a compound of sodium, oxygen, and chlorine). A gallon of household bleach contains 5.25 percent sodium hypochlorite and 94.75 percent water.
- Chlorine bleaches cotton well because both bleach and cotton are alkaline on the pH scale.
- To commercially prepare wool for bleaching, manufacturers soak the wool in acid before and after the bleaching process to neutralize the pH level and prevent the bleach from dissolving the wool.
- Never mix any acidic liquid with chlorine bleach. Doing so can cause a chemical reaction that releases toxic gases that can kill you.
- Like wool, human hair is also acidic. Substitute a lock of hair for the wool in this experiment to observe its reaction with chlorine bleach.

Chlorine Keeps It Clean

- Chlorine is a yellow-green gas, but its common compounds—known as chlorides—are usually colorless.
- The name *chlorine* is from the Greek word *chloros*, meaning "greenish yellow."
- The most common compound of chlorine is sodium chloride, commonly known as salt.
- Swedish chemist Carl Wilhelm Scheele discovered chlorine in 1774, convinced it was a compound that contained oxygen. In 1810, English chemist Sir Humphry Davy proved that chlorine was actually an element.
- Used in drinking water and swimming pool water to kill harmful bacteria, chlorine is also used as part of the sanitation process for industrial waste and sewage.
- During World War I, chlorine was used in chemical warfare as a poisonous gas. Heavier than air, chlorine gas stays close to the ground, spreads rapidly, and fills low-lying foxholes and trenches.
- Household chlorine bleach can release deadly chlorine gas if it is mixed with certain other cleaning agents, such as ammonia.

POTASSIUM

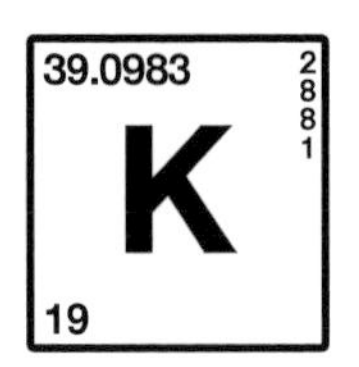

31. Daring Diet Coke and Mentos Rocket

SAFETY FIRST

This experiment should be performed outside.

- Safety goggles

WHAT YOU NEED

- 2-liter soda bottle, clean and empty
- Utility knife or scissors
- Mentos mints
- Nail, 1¾-inch 5d finish
- Needle-nose pliers
- Small metal paper clip
- Cork, 1 inch by ¼ inch (available at craft stores)
- Full 2-liter bottle of Diet Coke, sealed
- Hammer

WHAT TO DO

1. Remove the label from the clean, empty 2-liter soda bottle.
2. Using a utility knife or scissors, cut the soda bottle in half around the center of the bottle, and discard the top half.

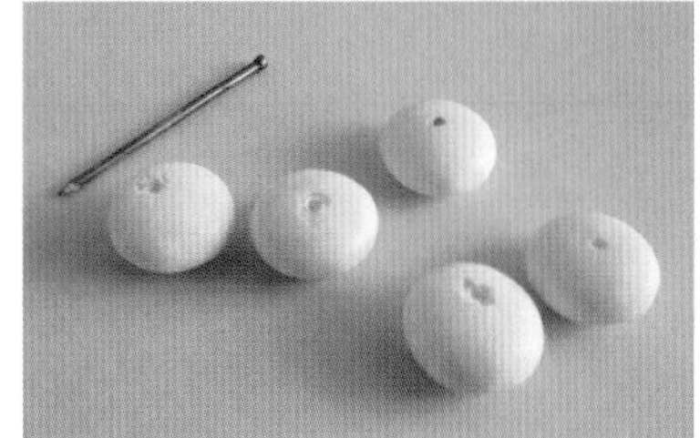

The remaining bottom half of the bottle will serve as your launchpad.

3. Open the pack of Mentos and take out five mint candies. Gently punch a hole through the center of each of the Mentos candies by slowly pushing the nail through it without cracking the hard candy coating.
4. Using a pair of needle-nose pliers, straighten the paper clip, leaving the smallest end loop intact, creating the shape of the letter *J*.
5. Thread the prepared Mentos candies onto the paper clip. The end loop will hold them in place.
6. Outside, insert the free end of the paperclip into the bottom center of the cork so that the candies are pressed firmly against the cork.

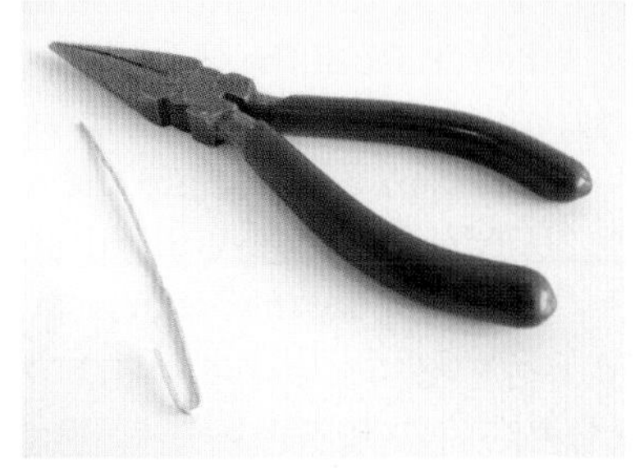

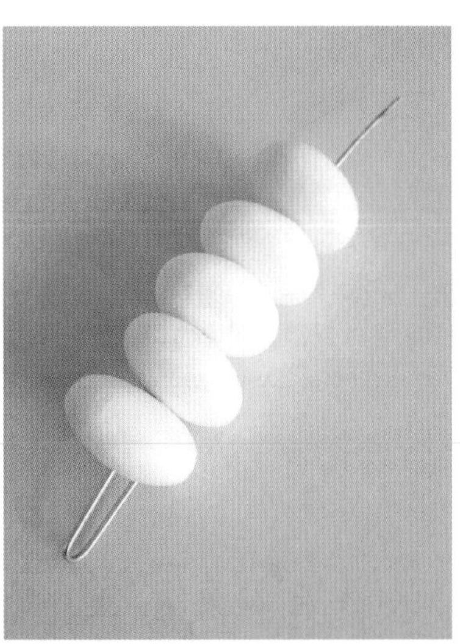

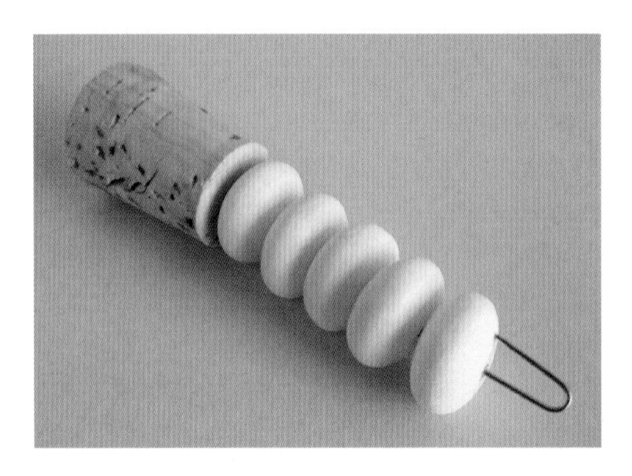

7. Open the full 2-liter bottle of Diet Coke and remove the cap.
8. Pour out some of the soda to leave just enough room so that when you insert the cork, the Mentos candies sit above the surface of the soda.
9. Wearing safety goggles, carefully insert the prepared cork into the neck of the bottle. Press the cork firmly into the neck, and use the hammer to secure the cork tightly. Do not insert the cork too tightly into the bottle, otherwise the pressure created by the reaction between the Diet Coke and the Mentos candies can cause the plastic bottle to explode.
10. Place the previously created launchpad (the bottom half of the first bottle) upside down on top of the corked Diet Coke bottle, capping the full bottle.
11. Quickly flip the device upside down, set it on level pavement, and step back quickly.

WHAT HAPPENS

The Diet Coke shoots the cork out from the neck of the upside-down bottle for approximately 10 seconds, sending the plastic bottle shooting anywhere from 15 to 40 feet in the air.

WHY IT WORKS

The potassium benzoate and aspartame in the Diet Coke reduce the work required for bubble formation, allowing carbon dioxide to rapidly escape from the soda. The gelatin and gum arabic from the dissolving Mentos candies interact with the potassium benzoate and aspartame in the diet soda, weakening the surface tension of the water in the soda and allowing the carbon dioxide bubbles to expand. At the same time, the rough surface of the Mentos candies lets new bubbles form more quickly (a process called nucleation). As more of the candies dissolve, both processes accelerate, rapidly producing foam. The resulting pressure inside the bottle pops the cork from the bottle's neck and forces a geyser of foam to spray from the bottle. As Sir Isaac Newton's third law of motion states: For every action, there is an equal and opposite reaction. When the soda shoots from the neck of the bottle, the reaction to it sends the soda bottle flying upward.

WACKY FACTS

- The basic principle behind this Diet Coke and Mentos Rocket is exactly the same principle behind a rocket launching into space. When the fuel burns, gas escapes from the rocket's bottom, and the opposite reaction sends the rocket upward.
- Diet Coke works better than regular Coca-Cola as fuel for the rocket because the aspartame in the diet soda lowers the surface tension of the liquid more than the sugar or corn syrup in the regular variety.
- On January 13, 1920, the *New York Times* editorialized that rocket scientist Robert H. Goddard "does not know the relation of action to reaction, and of the need to have something better than a vacuum against which to react—to say that would be absurd. Of course he only seems to lack the knowledge ladled out daily in high schools." Five years later, Goddard launched the first liquid-fuel rocket. On July 17, 1969, three days before the *Apollo 11* astronauts landed on the moon, the *New York Times* retracted the statement.
- You can make a car powered by Diet Coke and Mentos by strapping the prepared soda bottle (in step 8 above) to the top of a skateboard with large rubber bands or duct tape and then placing the skateboard on pavement.

•

32. Preposterous Purple Blaze

SAFETY FIRST

This experiment should be performed outside.

- Safety goggles
- Rubber gloves

WHAT YOU NEED

- Measuring cup
- Hot water
- Plastic bucket
- Morton Salt Substitute (3.125 ounces) or NoSalt (11 ounces—both available at grocery stores)
- Pinecones (open, not closed)
- Small brick or paver stone
- Mesh laundry bag (optional)
- Fireplace or campfire

WHAT TO DO

1. Wearing safety goggles and rubber gloves, pour 2 quarts of hot water into the plastic bucket.
2. Add 3 ounces of Morton Salt Substitute or NoSalt, and swirl the bucket until the salt substitute (potassium chloride) dissolves in the water.
3. Submerge the dry pinecones in the solution (no more than the solution can cover completely), and place the small brick or paver stone on top of the pinecones to hold them underwater.
4. Let the pinecones soak in the solution overnight.
5. Remove the pinecones and let dry on a flat surface (or hang in a mesh laundry bag) for 3 days.
6. Carefully add one pinecone at a time to the glowing orange embers of your fireplace or campfire and observe.

WHAT HAPPENS

The glowing orange embers ignite the pinecone, which burns with a purple flame.

WHY IT WORKS

In 1777, French chemist Antoine Laurent Lavoisier proved that fire is the heat and light that results from the rapid union (or combustion) of oxygen with other substances. The color of the flame depends on the substance being united with the oxygen. In this experiment, the pinecone provides the fuel and kindling to ignite the potassium chloride (a compound of potassium and chlorine), which gives the flame a violet color.

WACKY FACTS

- Other chemicals produce colored flames. A pinecone soaked in a half gallon of water and three ounces of salt (sodium chloride) produces yellow flames. A pinecone saturated in a solution made from a half gallon of water and three ounces of 20 Mule Team Borax will burn green, and a pinecone immersed in a half gallon of water and three ounces of Epsom salt will yield white flames.

- Ever since the discovery of fire, people have tried to figure out how to harness the energy of fire to create more light. They discovered that wood dipped in pitch would burn brighter and longer. Eventually, people discovered that a wick placed in a bowl of oil would burn brightly.
- Scientists use flame tests to identify the elements present in a chemical. They grind up the chemical sample, mix it with methylated spirit, and light it with a match. The color of the flame helps indicate which type of salts the chemical contains.

Go Bananas for Potassium

- When exposed to water, potassium, a highly reactive metal, explodes with a purple flame.
- Although potassium is the seventh most abundant element in the Earth's crust, the element is not found free in nature but rather in minerals and compounds, such as saltpeter and potash.
- In 1807, English chemist Sir Humphry Davy isolated potassium by running an electric current through some wet potash (potassium carbonate).
- Potassium is named after the salt potash. The atomic symbol K stems from the Latin word *kalium*, meaning "potash."

CALCIUM

40.078
2
8
8
2
Ca
20

33. Silly Naked Eggs

WHAT YOU NEED

- 24-ounce glass jar with lid, clean and empty
- Vinegar, 5 percent acetic acid
- 1 raw egg
- Water
- Red food coloring

WHAT TO DO

1. Fill the glass jar with vinegar, carefully drop in the egg, and seal the lid.

2. Let the jar sit undisturbed for 2 days or until the eggshell completely disintegrates, leaving only the membrane.
3. Carefully cover the mouth of the jar with your hand and pour the liquid through your fingers and down a sink drain, catching the egg in your hand and being careful not to poke the delicate membrane.
4. If any egg shell remains, gently rub it off.
5. Fill the jar with water, add 10 drops of red food coloring, reseal the lid, and let sit for 24 hours.
6. Observe the egg, and let sit for several more days.

WHAT HAPPENS

Shortly after you pour vinegar into the glass, small bubbles appear on the eggshell, and after 24 hours, a frothy white layer of scum appears on the surface of the vinegar. After two days, the egg has no shell, feels rubbery, and appears slightly larger than it was originally. When left in the red water, the egg turns red and enlarges until the membrane bursts.

WHY IT WORKS

An eggshell is composed primarily of calcium carbonate (a compound of calcium, carbon, and oxygen). When you soak the egg in vinegar, the acetic acid dissolves the calcium carbonate, releasing carbon dioxide gas, which you see as bubbles on the eggshell. The fragile membrane beneath the eggshell remains intact and holds the inside of the egg (the albumen and yolk) together. The egg appears slightly larger because some of the vinegar migrates through the membranes to the inside of the egg through osmosis. The red water, being a less concentrated solution than the egg's contents, passes through the semipermeable egg membrane, turning the egg red and eventually causing the pressure inside the egg to increase until the membrane bursts.

WACKY FACTS

- The rubbery egg will actually bounce if dropped from not too great a height. Experiment by dropping the egg from one inch above a tabletop, increasing to two or three inches.
- To shrink the egg (before it bursts), fill a jar with a few inches of corn syrup and carefully place the naked egg in the jar. Let sit for two or three days. Corn syrup contains a small amount of water, so the water inside the egg will migrate through the semipermeable egg membrane and into the corn syrup through osmosis to equalize the water concentration. Eventually the egg will look like a shriveled prune that contains a yoke.
- To revive the shriveled egg, place the egg in a jar of water for several days. The water will migrate through the membrane again through osmosis and refill the egg with water.
- A naked egg is not an edible pickled egg. Pickled eggs are shelled hard-boiled eggs soaked in brine made from vinegar.

Calcium: Feel It in Your Bones

- Calcium occurs in nature only in compounds, most notably calcium carbonate, calcium fluoride, and calcium sulfate.
- The ancient Egyptians, Greeks, and Romans used calcium compounds to make mortar.
- English chemist Sir Humphry Davy first isolated calcium as a pure metal in 1808. He named calcium after the Latin word *calcis*, meaning "lime." The Romans extracted calcium from limestone, marble, and chalk.
- Calcium is also a vital element for both plant and animal life. Calcium makes bones and teeth hard. The best sources of calcium for the human body include cheese, yogurt, milk, salmon, and tofu.

34. Weird Walk on Eggs

WHAT YOU NEED

- Pencil
- Straightedge ruler
- Utility knife
- Foam-core poster board
- Duct tape
- Scissors
- 4 cartons of eggs (4 dozen eggs total)

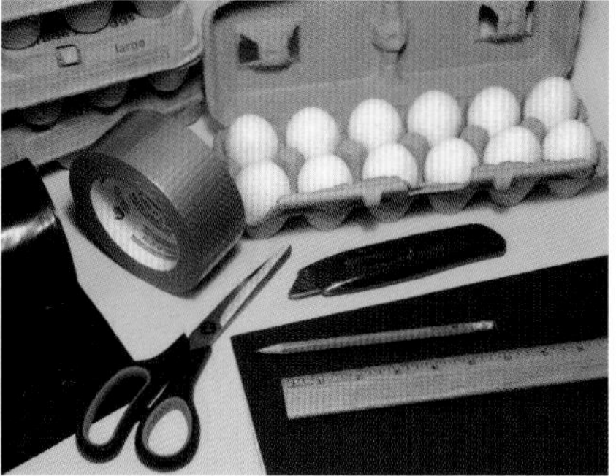

WHAT TO DO

1. Use the pencil, straightedge ruler, and utility knife to carefully measure and cut two 14-inch-by-6-inch rectangles from the foam-core poster board.
2. Tear off a 10-inch length of duct tape, and fold it in half

lengthwise (sticking the tape to itself).

3. Place your bare foot in the center of one of the rectangles of foam-core poster board, place the prepared strip of duct tape across the top of your foot like a sandal strap, and use additional strips of duct tape to adhere each end of the strap to the underside of the poster board.
4. Repeat step 3 with your other foot and the second rectangle of poster board.

5. Using scissors, carefully cut off the tops of the egg cartons.
6. Place the four egg cartons on the floor, making certain all the eggs are positioned pointy-side up in the cartons.
7. Wearing the foam-core poster board sandals on your bare feet, slowly step on top of the eggs.

WHAT HAPPENS

The eggs support your weight and do not crush or break.

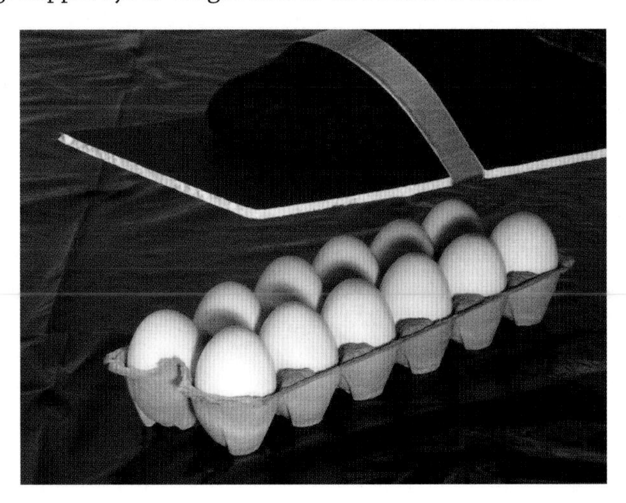

WHY IT WORKS

The shell of each egg is made up of 95 percent calcium carbonate (a compound of calcium, carbon, and oxygen) and a flexible protein matrix, and the arch shape along the longitudinal axis of each egg provides remarkable strength. Pressure applied to the top of the egg is uniformly distributed around the egg, so no one particular spot on the egg bears the load. An individual egg can bear roughly 33 pounds of pressure before the shell breaks. A dozen eggs standing upright in a carton can collectively hold roughly 400 pounds of pressure. The sheet of sturdy foam-core poster board helps distribute the weight put on each foot equally across all the eggs underneath.

WACKY FACTS

- Aside from 95 percent calcium carbonate, an eggshell is composed of 5 percent other minerals, including calcium phosphate and magnesium carbonate.
- The average ostrich egg, approximately 24 times the size of a hen's egg, can support the weight of a 280-pound human.
- Eggs do not crush under the weight of a mother bird as she sits on the nest because when a force is applied to an egg, the curve of the egg distributes the force over a wide area away from the point of contact.
- If you hold an egg in your palm and then try to squeeze your hand into a fist, you will not crush the egg. Similarly, if you hold the egg between your thumb and index finger (with the bottom of the egg held by your thumb and the top of the egg held by your index finger), you will not be able to crush the egg by squeezing your thumb and index finger together.
- Newspaper editor Joseph L. Coyle of Smithers, British Columbia, invented the egg carton in 1911 to end a dispute. A local farmer was shipping eggs to a hotel in Aldermere, British Columbia, and the eggs frequently arrived broken. To stop the farmer and hotel owner from blaming each other, Coyle designed an egg carton made from paper to protect the eggs during their journey from the farm to the hotel.

IT'S ELEMENTARY

Playing with Plutonium

In 1940, four American nuclear scientists in Berkeley, California, decided to bombard uranium with nuclei. The result? They artificially created the element plutonium, a material twice as dense as lead that gives off radiation in the form of alpha rays, making it highly poisonous and carcinogenic.

Plutonium is also highly explosive. In quantities larger than 10 kilograms, the isotope plutonium-239 explodes spontaneously. Scientists used plutonium-239 in the second atomic bomb, which was dropped on the Japanese city of Nagasaki in 1945, killing or injuring more than 200,000 people and effectively ending World War II. Nuclear reactors that use uranium as a basic fuel create plutonium-239 as a waste product. Unfortunately, the half-life of plutonium-239 (the time required for 50 percent of its radioactivity to disappear through decay) is 24,100 years, making the disposal of plutonium-239 a serious problem.

The isotope plutonium-238, created as a by-product of making bomb-grade plutonium-239, has a half-life of just less than 88 years. Scientists have used plutonium-238 as a battery to power spacecrafts for decades, including *Voyager 1* (which left Earth in 1977 to explore Jupiter and Saturn and reached interstellar space in 2014), *Voyager 2*, *Cassini* (in orbit around Saturn), *Galileo* (which traveled to Jupiter), *Curiosity* (exploring the surface of Mars), and *New Horizons* (which flew by Pluto in 2015). SNAP-27, a plutonium-238-fueled battery, powered the science experiments of every lunar landing after *Apollo 11* and still generates power on the moon.

47.867 2 8 10 2
Ti
22

35. Mysterious Memory Metal

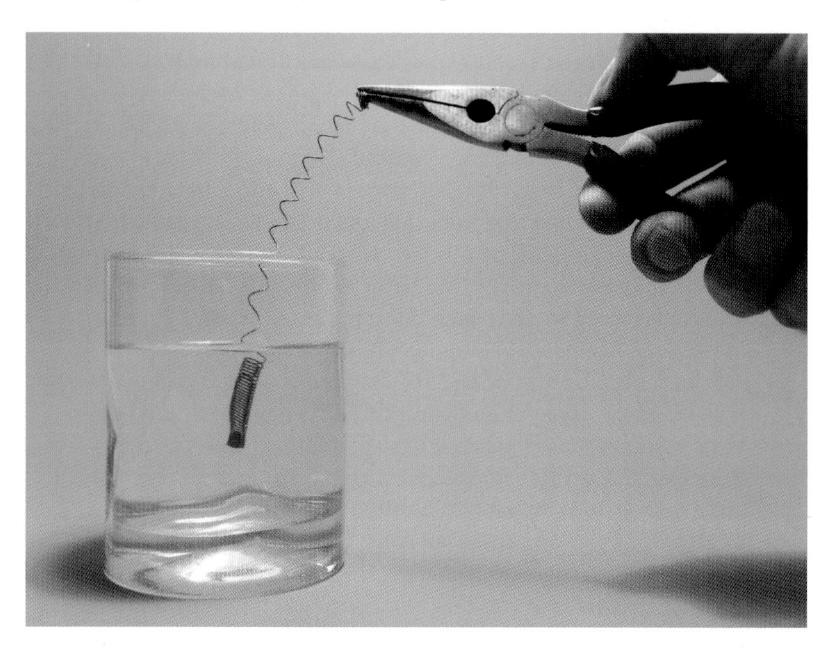

WHAT YOU NEED

- Nitinol helical spring—standard transition temperature (115°F), 3/16-inch (4.7-millimeter) mandrel size, ½-millimeter pitch, continuous cutting (available at Kellogg's Research Labs, www.kelloggsresearchlabs.com)
- Needle-nose pliers
- Drinking glass filled with hot water
- Wire cutters
- Two lengths of electrical wire (22 gauge), 12 inches long
- 9-volt battery

WHAT TO DO

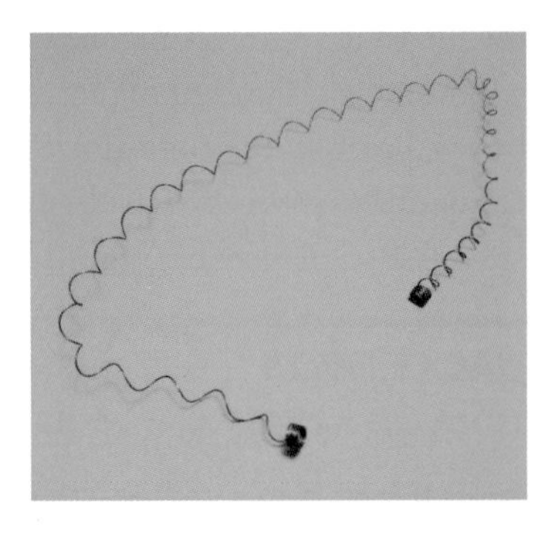

1. With the spring at room temperature, grab both ends of the spring and stretch it out.
2. Try to coil the spring back together to regain the original shape.
3. Holding one end of the stretched spring with a pair of needle-nose pliers, slowly submerge the spring in the glass of hot water.
4. Remove the spring from the water.
5. Stretch the spring again.
6. Use the wire cutters to strip 1 inch of plastic coating off each end of the two 12-inch lengths of electrical wire.

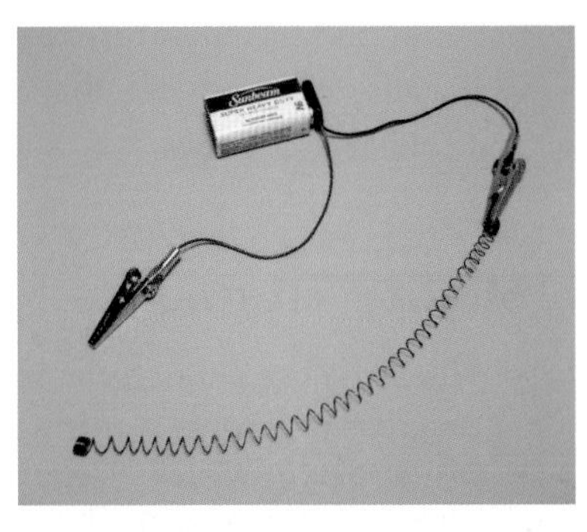

7. Wrap one end of the first wire to one end of the spring.
8. Wrap one end of the second wire to the other end of the spring.
9. Touch the free end of one wire to the positive end of the 9-volt battery.
10. Touch the free end of the second wire to the negative end of the battery.

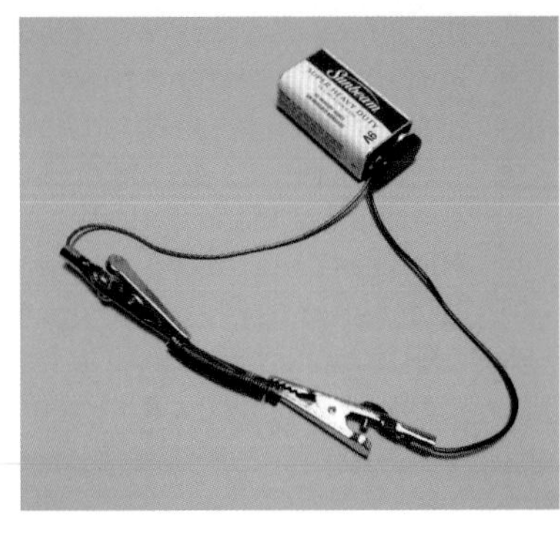

WHAT HAPPENS

When dipped in the hot water, the stretched spring recoils to its original shape. When wired to the battery, the stretched spring returns to its original shape.

WHY IT WORKS

Nitinol, an alloy composed of two metals (nickel and titanium), was discovered at the Naval Ordnance Laboratory in White Oak, Maryland. Nitinol possesses both superelasticity and shape memory. Superelasticity means that an object made from nitinol can be twisted or bent only at a "transformation temperature" range above a certain temperature and below a certain temperature. Shape memory means that an object made from nitinol returns to its original shape at a temperature lower or higher than the transformation temperature range.

WACKY FACTS

- Nitinol gets its name from the atomic symbols for nickel (Ni) and titanium (Ti), combined with the abbreviation for Naval Ordnance Laboratory (NOL).
- Nitinol springs are used in antiscalding faucets, so when water rises to too high a temperature, the spring will recoil, shutting off the faucet to prevent injury.
- Doctors insert nitinol stents into clogged arteries, allowing the patient's body temperature to expand the stent.
- Other alloys that possess shape memory include gold-cadmium alloys and copper-zinc alloys.

Mighty Titanium

- Titanium does not occur naturally in its pure form but bonded together with other elements.
- British pastor William Gregor discovered titanium in 1791. German chemist Martin Heinrich Klaproth named titanium after the Titans of Greek mythology.
- In 1910, New Zealand chemist Matthew A. Hunter isolated titanium, using a method now called the Hunter process.
- Titanium resists corrosion (including in sea water and chlorine) and has the highest strength-to-weight ratio of any metal.
- Titanium is as strong as many steels yet 45 percent lighter.
- Titanium is used as a component in laptop computers, firearms, sports equipment, bicycle frames, camping cookware and utensils, and surgical implements and implants.

36. Cool Colors

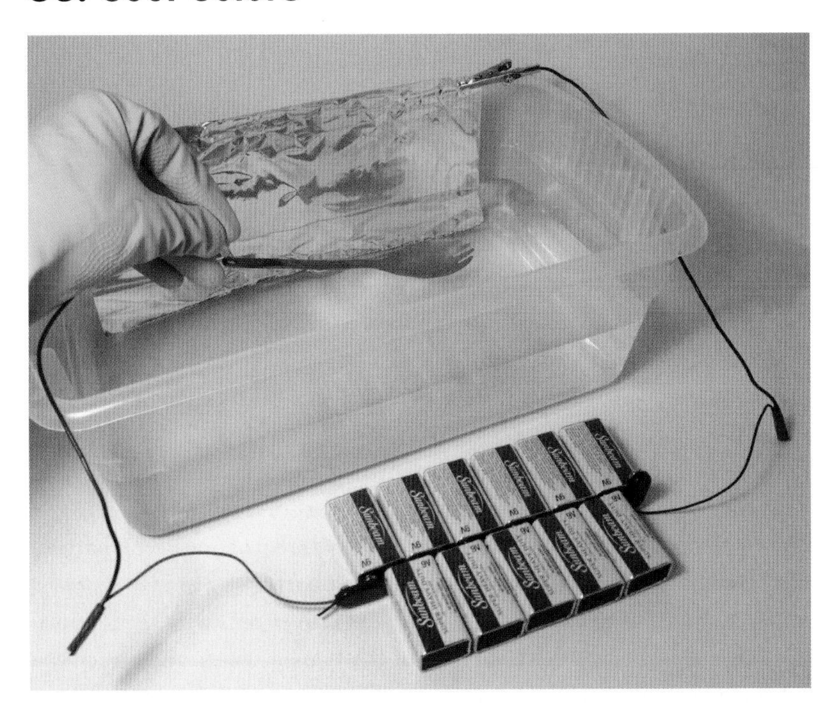

SAFETY FIRST

- Rubber gloves

WHAT YOU NEED

- Wire cutters
- 2 battery snap connectors (available at RadioShack), 9 volt
- 2 lengths of electrical wire (22 gauge), 12 inches long
- 2 alligator clips
- Electrical tape
- Isopropyl alcohol, 70 percent
- Paper towels
- Titanium camping spork (available at sporting goods stores)
- 9 batteries, 9 volt

- Aluminum foil sheet, 12-by-12-inch square
- Plastic shoebox container
- Vinegar, 5 percent acetic acid

WHAT TO DO

1. Use the wire cutters to snip off the positive (red) wire on one battery snap connector and the negative (black) wire on the second battery snap connector.
2. Use the wire cutters to strip ½ inch of the insulation from both ends of the two 12-inch wires.
3. Attach an alligator clip to one end of each 12-inch wire.
4. Twist the free end of one 12-inch wire to the exposed end of the positive (red) wire from one battery snap, and twist the free end of the second 12-inch wire to the exposed end of the negative (black) wire from the second battery snap.
5. Wrap each set of the exposed, twisted wires with a strip of electrical tape.
6. Use isopropyl alcohol and a paper towel to clean any oils, dirt, and other residue from the titanium camping spork.
7. Wearing rubber gloves, attach the positive snap of one 9-volt battery into the negative snap of a second 9-volt battery.
8. Attach the snap connector with the positive (red) wire to the positive snap remaining free on one of the two batteries, and attach the snap connector with the negative (black) wire to the negative snap remaining free on the second battery. Do not allow the two alligator clips to touch each other.
9. Fold a 12-inch-square sheet of aluminum foil over one side of the plastic shoebox container and connect the alligator clip attached to the negative (black) wire on the battery snap connector to the foil.
10. Fill the plastic shoebox container with vinegar.
11. Connect the alligator clip attached to the positive (red) wire to the end of the handle of the titanium spork.
12. Without dipping the alligator clip into the vinegar, hold the titanium spork submerged for 30 seconds, approximately ¼ inch away

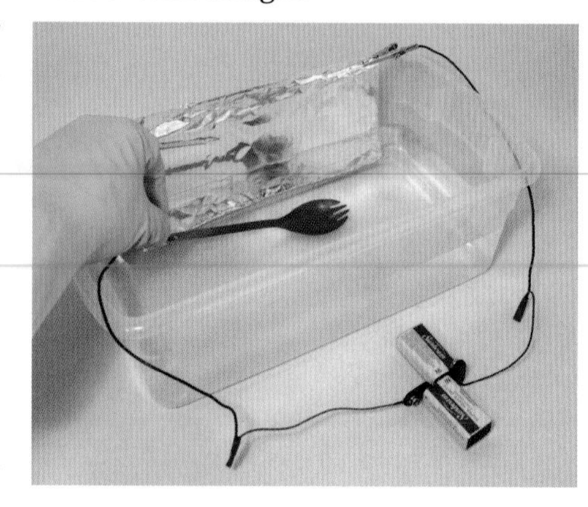

from the aluminum foil. Do not allow the titanium spork (anode) to touch the aluminum foil (cathode). Doing so may short-circuit the batteries. Tiny bubbles will form around the titanium and the aluminum foil. You may also hear fizzing and notice steam rising from the vinegar.

13. Remove the titanium spork from the vinegar, observe the color, reposition the alligator clip on the spork, and submerge the spork again for 30 seconds to color areas that were not previously submerged in the vinegar.
14. Remove the titanium spork from the vinegar, add another battery to the series of batteries, and repeat steps 12 through 14 until you have added all the batteries—one by one—to the power supply.

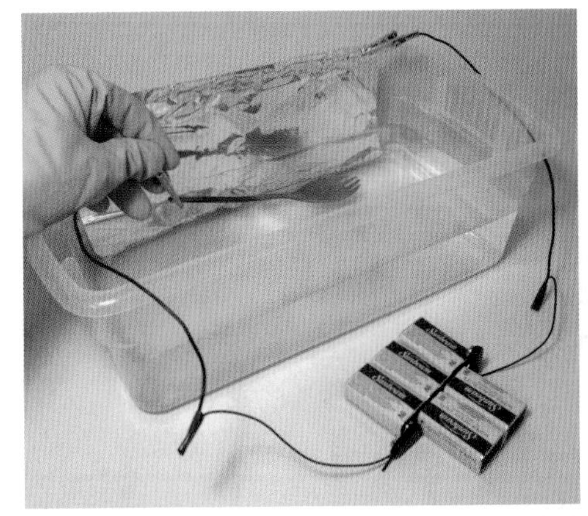

WHAT HAPPENS

The color of the titanium changes from silver through a sequence to bronze, purple, blue, yellow, pink, and turquoise.

WHY IT WORKS

Anodizing is a process of adding a coating on a metal. Titanium reacts when exposed to oxygen, forming a clear layer of titanium dioxide on the exposed surface of the titanium. When titanium serves as an anode (the positively charged electrode by which the electrons leave a battery) in a circuit, a thicker layer of oxide forms on the surface of the titanium. Connecting the batteries in series increases the voltage from

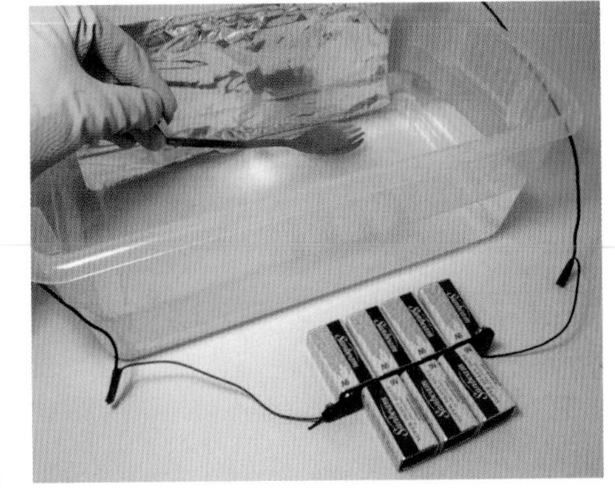

9 volts per battery to 81 volts with a total of nine 9-volt batteries. The sheet of aluminum foil serves as the positively charged cathode, and the vinegar serves as the electrolyte in this anodizing process. The greater the voltage, the thicker the layer of oxide formed on the titanium. The thickness of the layer of oxide interferes with light waves reflected off the metal surface, producing the resulting color.

WACKY FACTS

- The thickness of the oxide layer formed during the anodizing process depends on the voltage of electricity applied to it. You can use all nine batteries at once and remove the titanium object from the vinegar bath when the piece attains the color you desire. Or you can use the chart below to determine the number of batteries to achieve the color you wish.

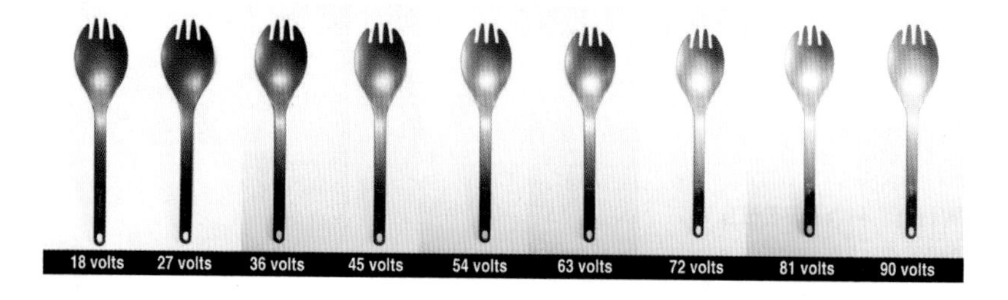

- The precise voltage required to attain a particular color depends on several variables, including the stability of the batteries, the free-ion content of the electrolyte, and the finish of the surface of the titanium piece. To anodize two pieces of titanium the exact same color, anodize them simultaneously.
- Other electrolytes that work well for anodizing titanium include Coca-Cola, borax (one teaspoon dissolved in four cups of water), or trisodium phosphate (one teaspoon dissolved in four cups of water).

CHROMIUM

37. Sizzling Sphere Smashup

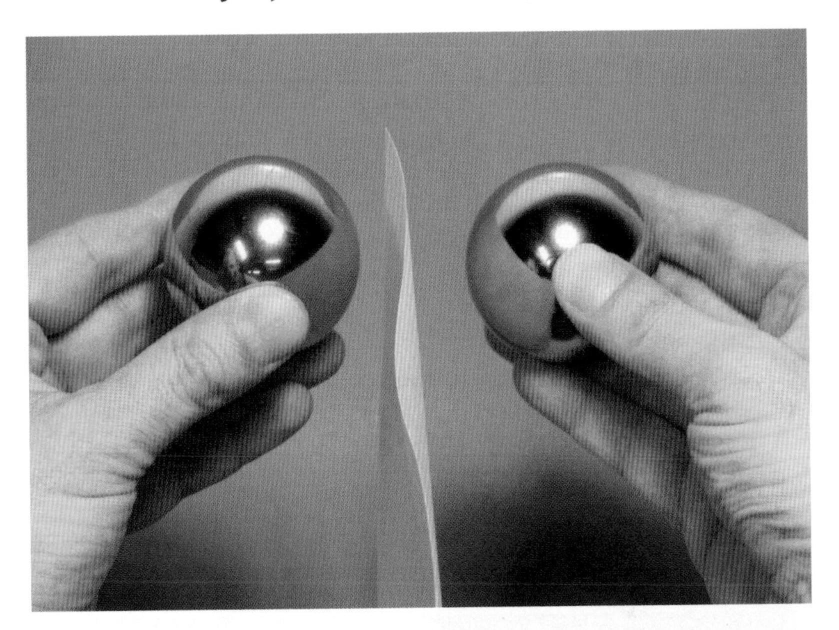

WHAT YOU NEED

- 2 chrome steel balls, 1 pound each, 2-inch diameter
- Sheet of white paper
- Sheet of aluminum foil

WHAT TO DO

1. Hold a chrome steel ball in each hand, position your hands approximately 3 feet apart, and quickly bring your hands together to bang the balls together.
2. Have an assistant hold the sheet of white paper vertically at the spot where you knocked the two balls together, so when you attempt

to knock the two balls together again, the piece of paper is struck between them.

3. Examine and smell the spot on the sheet of paper where the two chrome balls collided.
4. Have your assistant hold the sheet of aluminum foil vertically at the same spot, and knock the two chrome balls together again, striking the aluminum foil.
5. Examine the spot on the aluminum foil where the two chrome balls collided.

WHAT HAPPENS

The chrome-ball collision burns a small hole through the sheet of white paper, and you can smell smoke. At the spot where the chrome balls hit the sheet of aluminum foil, you can observe a number of concentric circles embedded in the foil.

WHY IT WORKS

When you smash the two chrome steel balls together, the kinetic energy (or the energy of motion) converts into thermal energy, generating

enough heat at the point of contact to burn a hole in a piece of paper. Many scientists believe that at the point of collision, a shock wave travels through the paper or foil, creating the concentric ripples in the foil.

WACKY FACTS

- Chrome spheres smaller than two inches in diameter do not have enough mass to generate sufficient heat to burn through the paper. Chrome spheres larger than two inches in diameter have too much surface area at the point of collision to generate sufficient heat to burn through the paper.
- Knocking two rusty spheres together with a piece of aluminum foil between them produces sparks.
- Some energy from the motion of the balls is converted into sound waves when the spheres collide, causing a cracking sound.
- When the balls collide, some energy may also remain in the balls, causing them to vibrate.

At Home with Chrome

- French chemist Nicolas L. Vauquelin discovered chromium in 1797 and isolated the metal shortly afterward. He named the element after the Greek word *chroma*, meaning "color," because chromium—sometimes called chrome—forms a variety of different colored compounds.
- Chromium is used in stainless steel and chromium plating for automobile bumpers, door handles, and trim.
- Chromium is not found as a free element in nature but in the form of ores, primarily chromite.
- Chromium is mined in Albania, Russia, South Africa, Turkey, and Zimbabwe.
- Chromium compounds give rubies their red color and emeralds their green color.

54.938045 2 8 13 2
Mn
25

MANGANESE

38. Groovy Genie in a Bottle

SAFETY FIRST

This experiment should be performed outside.

- Safety goggles
- Rubber gloves
- Respirator
- Garden hose

WHAT YOU NEED

- Electric drill with ¼-inch bit
- Cork
- Measuring cup
- Funnel
- Pool oxidizer, 27 percent hydrogen peroxide (such as Aqua Silk, available at pool supply stores)
- Wine bottle, clean and empty
- Toilet paper
- Measuring teaspoon
- Manganese dioxide, 15 g (available at chemical supply stores or www.homesciencetools.com)
- Thread
- Plastic drinking straw (optional)

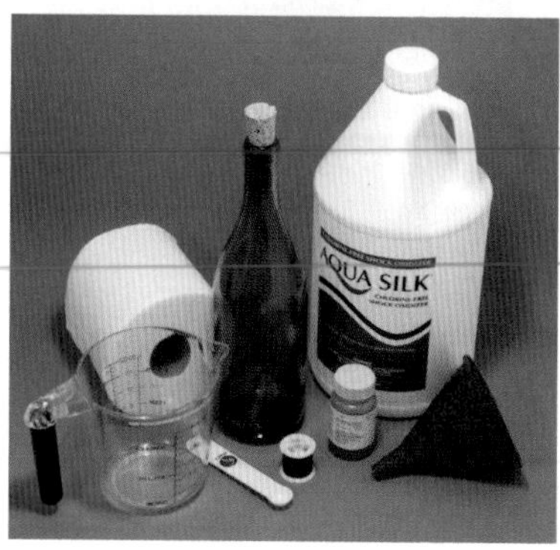

WHAT TO DO

1. Wearing safety goggles, drill a ¼-inch hole down through the center of the cork. (The hole is a safety precaution to allow vapor to escape from the wine bottle should you accidentally trigger the catalytic reaction.)

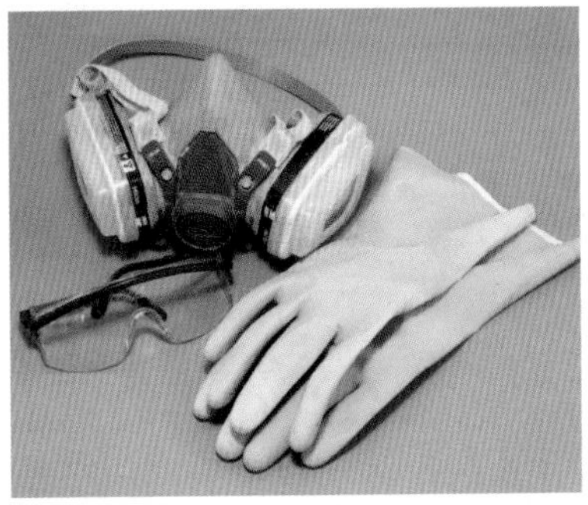

2. Wearing safety goggles, rubber gloves, and a respirator, use the measuring cup and funnel to carefully pour 3 ounces of pool oxidizer (27 percent hydrogen peroxide) into the wine bottle. **Note that 27 percent hydrogen peroxide is a strong oxidizer and is highly corrosive to eyes, skin, and the respiratory tract. Avoid all contact. In case of contact, flush with water for at least 15 minutes and seek medical attention if the eyes are affected. Do not mix 27 percent hydrogen peroxide with combustible materials.**
3. Tear off one square sheet of toilet paper, and place ½ teaspoon of manganese dioxide in the center of it. **Manganese dioxide is moderately toxic and a strong oxidizer. Contact with combustible or organic materials may cause fire.**

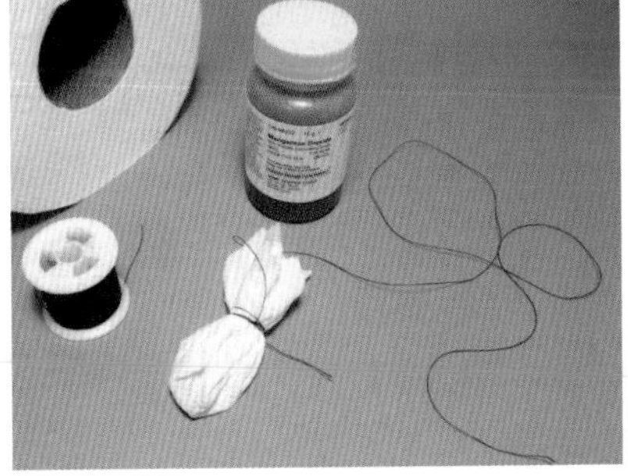

4. Fold up the corners of the square of toilet paper to form a small pouch.
5. Tie one end of the thread around the top of the pouch to secure it closed and to prevent the manganese dioxide from spilling out.
6. Holding the free end of the thread, carefully insert the toilet-paper pouch into the mouth of the wine bottle until the pouch hangs in the middle of the bottle. If necessary, use a plastic drinking straw to gently push the pouch through the

neck of the bottle. Do not let the pouch touch the hydrogen peroxide sitting at the bottom of the bottle.

7. Still holding the free end of the thread, replace the cork in the bottle so it holds the thread in place with the pouch hanging securely in the middle of the bottle.
8. Outside in a well-ventilated area with a garden hose accessible in case of an unforeseen chemical spill, set the bottle on a flat surface.
9. Remove the cork from the bottle, allowing the pouch of manganese dioxide to drop into the hydrogen peroxide, and step back.
10. When the reaction ceases completely and the glass bottle cools to the touch, empty the contents of the bottle down a sink drain and flush with excess water. Manganese dioxide is generally suitable for landfill disposal. Follow all federal, state, and local laws and regulations regarding the proper disposal of this chemical.

WHAT HAPPENS

A plume of steam spews from the bottle, as if a genie is being released, and the wine bottle heats up.

WHY IT WORKS

Hydrogen peroxide slowly decomposes into water and oxygen at room temperature. The manganese dioxide (a compound of manganese and oxygen) acts as a catalyst, speeding up the reaction and causing the hydrogen peroxide to rapidly decompose into water and oxygen gas in an exothermic reaction, releasing a great deal of heat, which changes the water into steam. The manganese dioxide is not consumed in the reaction and remains as a solid in the bottom of the bottle.

WACKY FACTS

- When nickel became scarce during World War II, the US mint replaced most of the nickel in its coins with manganese.

- The black color in cave paintings in Lascaux, France, was made 16,000 years ago using manganese dioxide.
- When hydrogen peroxide is used as a disinfectant for lacerations, the enzyme catalase present in blood catalyzes the breakdown of hydrogen peroxide into water and oxygen, causing bubbling.
- Despite its name, manganese is not magnetic.

Manganese Expertise

- The name manganese is derived from the Latin word *magnes*, meaning "magnet," because manganese compounds are used to add and remove color when making glass.
- In 1774, Swedish chemist Johann Gottlieb Gahn first isolated manganese from pyrolusite (now known as manganese dioxide).
- All animals and plants require a small amount of manganese to thrive. The average human body contains approximately 12 grams of manganese, found primarily in the skeleton.
- Manganese tarnishes when exposed to air, and the element rusts when exposed to moist air.

IRON

39. Bewildering Burning Steel

SAFETY FIRST

- This experiment should be performed outside.
- Safety goggles
- Metal tongs
- Fire extinguisher

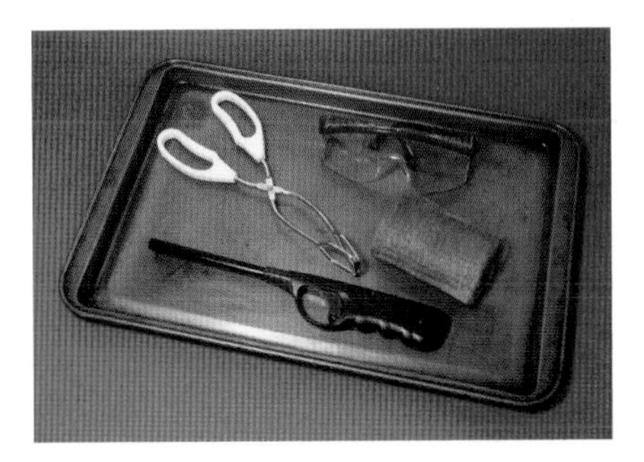

WHAT YOU NEED

- Steel wool pad, extrafine 0000 grade (available at hardware stores)
- Baking pan
- Butane barbecue lighter

WHAT TO DO

1. Wearing safety goggles and working outdoors in a well-ventilated area, gently pull apart the steel wool pad until it is the size and consistency of a cone of cotton candy.
2. Use the metal tongs to hold the wad of steel wool a few inches above the baking pan.
3. With a fire extinguisher on hand, use the butane barbecue lighter to ignite the steel wool.

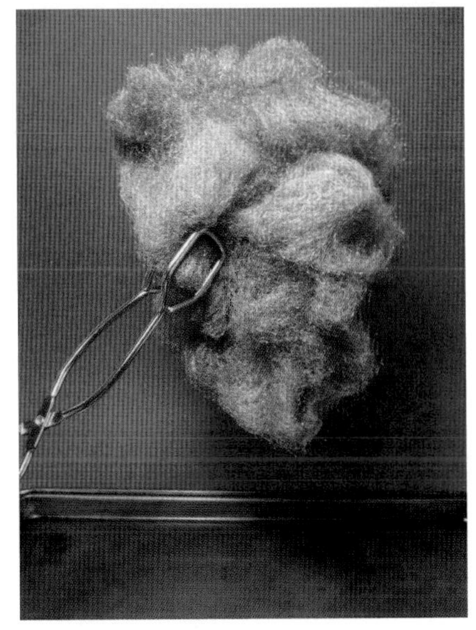

WHAT HAPPENS

The steel wool catches on fire, and the iron filings flash like a sparkler.

WHY IT WORKS

Combustion is the rapid chemical combination of a substance with oxygen, producing heat and light. The threads of iron in the steel wool, surrounded by more oxygen than in a solid block of iron, combust easily. For combustion to continue, the burning process must raise the temperature of nearby iron to its ignition temperature fast enough to sustain a chain reaction. Thick pieces of iron disseminate heat too fast to ever reach the ignition point. But thin strands of iron retain the heat and ignite nearby threads, quickly turning an entire steel wool pad into iron oxide.

WACKY FACTS

- The weight and mass of the steel wool in this experiment increases after it is burned due to the oxygen that combines with the iron.
- Rust (iron oxide) results when iron and oxygen react in the presence of water or moisture.
- Iron oxides are commonly used as pigments to color paint. Hematite (α-Fe_2O_3) is the most common iron oxide in red earth pigments, and the iron oxide hydroxide goethite (α-FeOOH) is the most frequently found iron compound in yellow earth pigments.

Fortified with Iron

- The atomic symbol for iron, Fe, comes from the Latin word *ferrum*, meaning "iron."
- Iron dissolves in water, but the process takes a long time.
- Iron is the fourth most common element in the Earth's crust.
- Adding a small amount of carbon to iron creates steel, which can be approximately 1,000 times stronger than pure iron.
- Cast iron is an alloy of iron that contains carbon, silicon, and manganese.
- The Iron Age, marking the widespread use of iron for tools and weapons, began between 1500 and 1000 BCE.
- The human body uses iron for a number of important functions, most notably in the red blood cells, where it forms an essential part of hemoglobin, a substance that carries oxygen from the lungs to other parts of the body.

58.6934 | 2 8 16 2

Ni

28

NICKEL

40. Gonzo Pickle Batteries

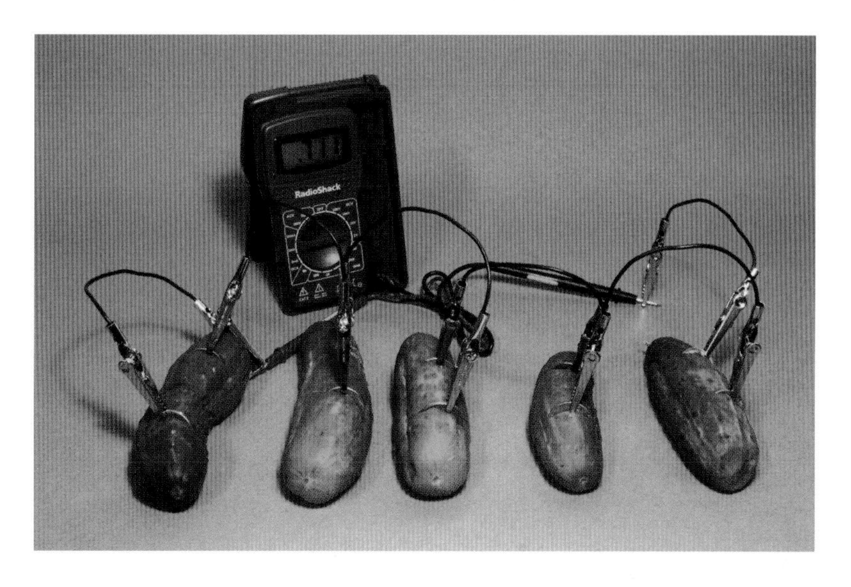

WHAT YOU NEED

- Wire cutters
- 6 lengths of electrical wire (22 gauge), each 12 inches long
- 12 alligator clips
- 5 kosher dill pickles
- Paper plate
- Sharp knife
- Ruler
- 5 shiny pennies (dated 1981 or earlier)
- 5 shiny nickels
- Multimeter (available at RadioShack)

WHAT TO DO

1. Use the wire cutters to strip ½ inch of the insulation from both ends of the six wires.
2. Connect an alligator clip to both ends of each wire.
3. Place a pickle on the paper plate, and using the knife, carefully cut two slits in the pickle. The slits should be 2 inches apart and face each another, perpendicular to the length of the pickle.

4. Insert a penny halfway into one slot and a nickel halfway into the other slot.
5. Repeat steps 3 and 4 for the remaining four pickles.
6. Pick up one wire and attach one alligator clip to the penny in the first pickle, and attach the remaining alligator clip from that same wire to the nickel in the second pickle.
7. Pick up a second wire and attach one alligator clip to the penny in the second pickle,

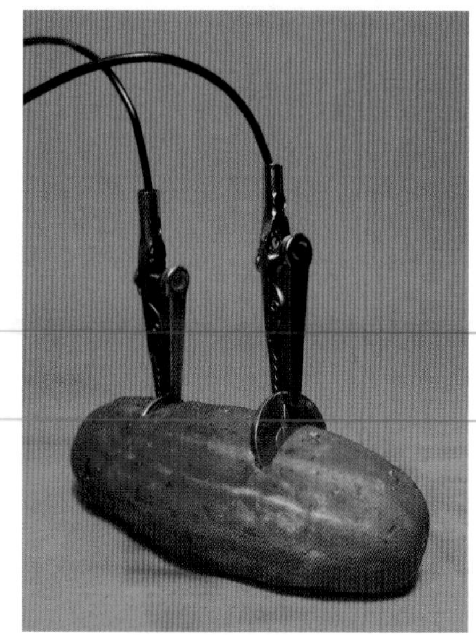

and then attach the remaining alligator clip from that same wire to the nickel in the third pickle.

8. Pick up a third wire and attach one alligator clip to the penny in the third pickle, and then attach the remaining alligator clip from that same wire to the nickel in the fourth pickle.
9. Pick up a fourth wire and attach one alligator clip to the penny in the fourth pickle, and then attach the remaining alligator clip from that same wire to the nickel in the fifth pickle.
10. Pick up a fifth wire and attach one alligator clip to the penny in the fifth pickle, and then attach the remaining alligator clip from that same wire to the black (negative) test probe on the multimeter.
11. Pick up a sixth wire and attach one alligator clip to the nickel in the first pickle and then attach the remaining alligator clip from that same wire to the red (positive) test probe on the multimeter.
12. Make certain that none of the alligator clips or pickles are touching each other. Also make sure that no pickles are sitting in a puddle of pickle juice from another pickle.
13. Set the selection dial on the multimeter on 2V in the DC current range, and observe the display on the multimeter for several minutes.

WHAT HAPPENS

The display on the multimeter shows that the five pickles, wired together in series like batteries, emit up to 0.7 volts of electricity.

WHY IT WORKS

The salty pickle juice acts as an electrolyte, conducting an electron flow between the copper in the penny and the nickel in the nickel, turning each pickle into a battery. The electrolyte causes the anode (the copper in the penny) to become negatively charged, and the cathode (the nickel in the nickel coin) to become positively charged.

WACKY FACTS

- Pennies minted in 1981 or before are 95 percent copper. Pennies minted in 1983 or afterward are 97.5 percent zinc with a thin copper coating. Pennies minted in 1982 could be either zinc or copper.
- The battery owes its discovery to frogs' legs. In the 1780s, Luigi Galvani, a professor of anatomy at the University of Bologna, noticed that the

legs of dead frogs twitched when they were hung from hooks on a rail. Fellow professor Count Alessandro Volta of the University of Pavia deduced that the frogs' legs were completing the circuit between the copper hooks and the iron rail, prompting him to produce a Voltaic cell in 1800.

- In the 1790s, Volta made the first battery by stacking pairs of silver and zinc disks separated from one another by cardboard disks moistened with a salt solution. The volt, a unit of electric measurement, is named after him.

The Fickle Nickel

- In medieval times, German copper miners in Saxony discovered a red mineral that looked like copper ore. Unable to extract copper from the rock, the miners named the ore *kupfernickel* after the German words *kupfer* (meaning "copper") and *nickel* (meaning "demon").
- In 1751, Swedish chemist Baron Axel Fredrik Cronstedt first identified nickel as a unique element while trying to extract copper from *kupfernickel*.
- Although nickel was commonly used to mint coins, the prevalence of allergic reactions to nickel through skin contact and the availability of cheaper metals has significantly reduced the widespread use of nickel in coinage.
- Nickel is used in magnets, rechargeable batteries, power tools, bathroom fixtures, and hybrid and electric vehicles.
- In the United States, a five-cent coin is called a nickel despite being composed of 75 percent copper and only 25 percent nickel.

63.546 2 8 18 1

Cu

29

COPPER

41. Ketchup and Copper

WHAT YOU NEED

- Ketchup
- Ceramic plate
- 12 brown pennies (not shiny)
- Water

WHAT TO DO

1. Pour the ketchup on the ceramic plate, creating a puddle approximately 6 inches in diameter.
2. Place the pennies (separated from each other) in the ketchup.
3. Pour more ketchup to cover the pennies.
4. Let sit undisturbed for 5 minutes.
5. Rinse the ketchup off the pennies with warm water.

WHAT HAPPENS

The ketchup removes the tarnish from the pennies, turning them bright and shiny.

WHY IT WORKS

When exposed to air, copper combines with oxygen to form copper oxide, which gives the penny a brown coating. When the penny is soaked in ketchup, the vinegar (acetic acid) in the ketchup reacts with the copper oxide to form copper acetate, which dissolves in water. The vinegar in the ketchup causes the salt (sodium chloride) in the ketchup to separate into sodium ions (charged atoms) and chlorine ions. The chlorine ions bond with the copper in solution forming copper chloride, allowing the acid to break more copper oxide free from the penny.

WACKY FACTS

- Do not conduct this experiment with pennies from a coin collection. The chemical reaction removes some of the copper from the penny, potentially effacing fine details and reducing the value of the coin.
- The Statue of Liberty is constructed from more than 89 tons of copper that is a mere 3⁄32 inches thick, the thickness of two pennies held together.

Get Wired with Copper

- The word *copper* and its atomic symbol Cu are derived from the Latin word *Cuprum*, meaning "Cyprus," the place where ancient Romans mined copper.
- The color of pure copper is reddish orange. When exposed to air, copper combines with oxygen to form copper oxide, darkening to a brown color. When exposed to air and water, copper forms a bright bluish-green encrustation called verdigris (primarily copper carbonate or copper chloride).

- Copper is an essential nutrient to all living organisms. Foods rich in copper include oysters, kale, mushrooms, seeds, nuts, beans, dried fruit, avocadoes, and goat cheese.
- Popular alloys of copper include brass (a mixture of copper and zinc) and bronze (an alloy of copper and tin).
- Used primarily in electrical wiring, copper is also a natural antibacterial. To prevent the spread of bacteria, brass doorknobs and handrails are frequently used in public buildings.

IT'S ELEMENTARY
Totally Sick Arsenic

As early as 500 BCE people knew about arsenic and its poisonous properties. The name *arsenic* stems from the Greek word *arsenikos*, meaning "potent." First identified as a metallic element by German alchemist Albertus Magnus in 1250 CE, arsenic, occurring free in nature as gray, brittle flakes, is a deadly poison used in insecticides and rat poison. Arsenic is commonly found in the minerals arsenolite, arsenopyrite, orpiment, and realgar.

Most arsenic compounds are odorless and tasteless, and when mixed in food or drink, provoke violent abdominal cramping, diarrhea, and vomiting—symptoms that mimic food poisoning and other common disorders, making arsenic poisoning difficult to detect. Death from shock follows in a few hours. As a poison, arsenic can be administered in a single large dose or in a series of smaller doses. The fatal dose of white arsenic (arsenic trioxide) is the size of a pea. The criminal use of arsenic as a poison to commit murder was common until 1836 when British chemist James Marsh developed chemical methods to detect arsenic poisoning.

In the 19th and early 20th centuries, doctors commonly used arsenic to treat syphilis, psoriasis, asthma, typhus, malaria, and menstrual cramps. Arsenic also conducts electricity easily, and gallium arsenide can produce laser light directly from electricity.

ZINC

65.38	2 8 18 2
Zn	
30	

42. Shocking Silver Pennies

WHAT YOU NEED

- Metal file
- Zinc screw (⅛ inch by 2 inches)
- Measuring cup
- Vinegar, 5 percent acetic acid
- 24-ounce glass jar, clean and empty
- Measuring teaspoon
- Epsom salt
- Sugar
- 1 cotton ball
- Isopropyl alcohol, 70 percent
- 3 clean, shiny pennies (dated 1981 or before)
- Wire cutters
- 2 lengths of electrical wire (22 gauge), each 12 inches long
- 4 alligator clips
- 6-volt lantern battery
- 2 plastic chip-bag clips
- Steel wool pad, extrafine 0000 grade (available at hardware stores)

WHAT TO DO

1. Using the metal file, carefully file the zinc screw to produce a small pile of powdered zinc large enough to cover this circle:

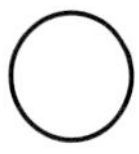

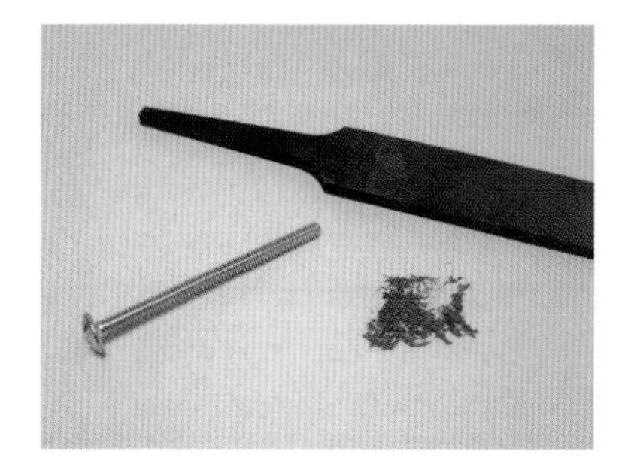

2. Pour ½ cup of vinegar into the glass jar. Add 1 teaspoon of Epsom salt and 1 teaspoon of sugar, and swirl the jar to dissolve.
3. Add the zinc powder to the solution, swirl, and let sit for 24 hours.

4. Saturate the cotton ball with 70 percent isopropyl alcohol and use it to clean the pennies, removing any oils, grime, and dirt. (If necessary, first use a clean, used toothbrush, baking soda, and water to scrub the pennies clean.)
5. Using the wire cutters, carefully strip ½ inch of insulation from both ends of the two lengths of electrical wire.
6. Connect an alligator clip to each of the four ends of the two wires.
7. Pick up one wire and attach one alligator clip to one end of the zinc screw. Attach the remaining alligator clip from that same wire to the positive battery terminal (anode).
8. Without allowing the alligator clip to touch the vinegar, submerge the zinc screw in the

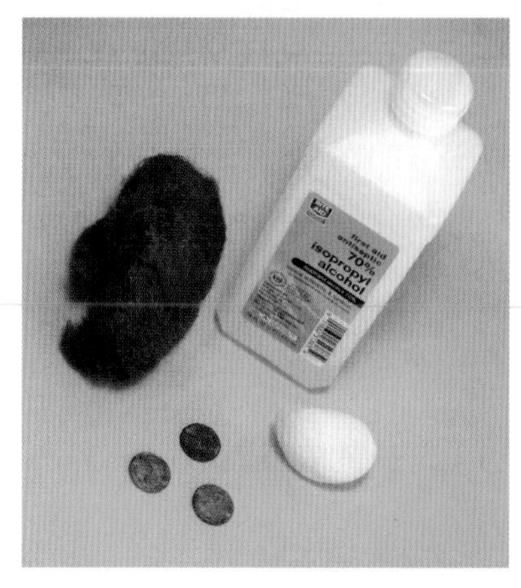

vinegar solution inside the jar. Use a plastic chip-bag clip to hold the wire in place against the rim of the jar.

9. Pick up the second wire and attach one alligator clip to a penny. Attach the remaining alligator clip from that same wire to the negative battery terminal (cathode).
10. Without allowing the alligator clip holding the penny (or the penny itself) to touch the vinegar, the zinc screw, or the screw's alligator clip, submerge the penny in the vinegar solution inside the jar. Use a plastic chip-bag clip to hold the wire in place against the rim of the jar opposite the wire connected to the zinc screw.

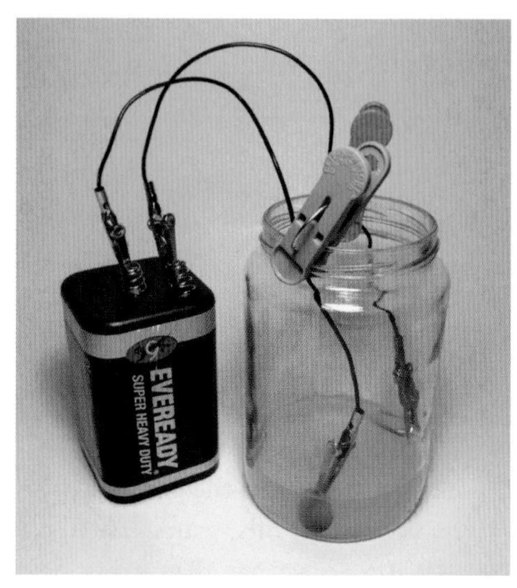

11. Wait 5 to 10 minutes.
12. Disconnect the alligator clips from the battery, and remove the penny from the solution. If the penny is a blackish color, buff it with the steel wool pad.
13. Repeat steps 9 through 12 with the other two pennies.

WHAT HAPPENS

The pennies appear to change from copper to silver.

WHY IT WORKS

The zinc powder reacts with the acetic acid in the vinegar to form hydrogen gas and zinc acetate. The teaspoon of Epsom salt (magnesium sulfate) makes the vinegar a better electrolyte, helping the acetic acid conduct electricity better. The teaspoon of sugar prevents the Epsom salt in the solution from crystallizing. The electric current from the battery carries the zinc molecules from the positively charged zinc screw and the molecules of zinc acetate to the negatively charged penny. When the zinc touches the surface of the copper penny, it turns to metallic zinc, coating the penny with a protective zinc layer that looks like silver, in a process called galvanization.

WACKY FACTS

- The composition of brass varies from 55 percent copper to more than 95 percent cooper and an inverse proportion of zinc.
- Because brass has a relatively low melting point (1,700°F), heating a penny too long can destroy the coating.

Think About Zinc

- In 1746, German chemist Andreas Marggraf first isolated zinc by heating carbon and calamine (yes, the active ingredient in calamine lotion).
- According to the International Zinc Association, zinc metal does not appear to have become available on a commercial scale until the 14th century, and Europeans did not recognize zinc as a separate metal until the 16th century.
- Half of all zinc produced today is used to galvanize steel to prevent iron and steel from rusting. Zinc is also used in batteries, electrical components, and household fixtures.
- Zinc oxide is used in rubber manufacturing and as a protective skin ointment.
- Zinc is an essential nutritional mineral for humans, helping the immune system fight off invading bacteria and viruses.
- Foods containing zinc include lean red meats, oysters, and peanuts.
- US pennies minted after 1983 are 97.5 percent zinc by weight, and when ingested by humans or animals cause the creation of toxic zinc salts in the acidic environment of the stomach.

107.8682

2
8
18
18
1

Ag

47

SILVER

43. Strange Silverware

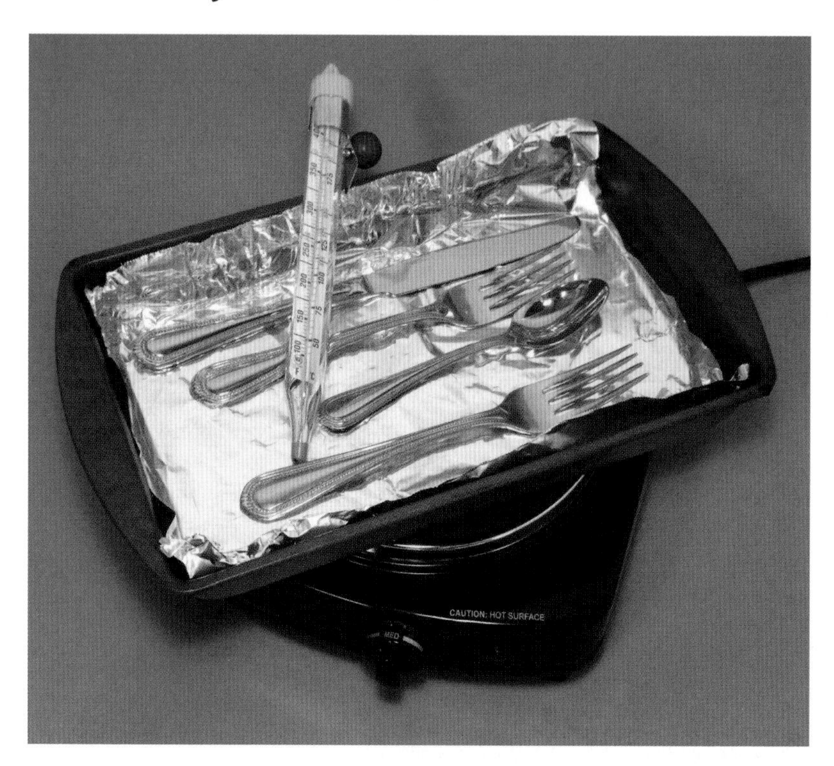

WHAT YOU NEED

- Metal cake pan
- Aluminum foil
- Measuring cup
- Water
- Measuring tablespoon
- Baking soda
- Cooking thermometer
- Hot plate or stove
- Tarnished silver

WHAT TO DO

1. Line the bottom of the metal cake pan with a sheet of aluminum foil, measure out enough water to cover the silverware (roughly 2 inches), and fill the cake pan with the water.
2. Add 2 tablespoons of baking soda per quart of water.
3. Using the cooking thermometer, heat the water above 150°F on a hot plate or stove.
4. Place the tarnished silverware in the pan so it rests on the aluminum foil. Do not let the water boil. Let the silverware soak over the heat for 10 minutes, then turn off the heat. Let the water cool before removing the silverware.

WHAT HAPPENS

The silver comes out sparkling clean—without any scrubbing.

WHY IT WORKS

Silver gradually darkens because silver chemically reacts with sulfur-containing substances in the air to form silver sulfide, a layer of black tarnish that coats the surface of silver. In this experiment, a chemical reaction converts the silver sulfide back into silver—without removing any of the silver. Like silver, aluminum forms compounds with sulfur—but with a greater affinity than silver. The warm baking soda solution carries the sulfur atoms from the silver to the aluminum, creating aluminum sulfide, which adheres to the aluminum foil or forms tiny yellow flakes in the bottom of the pan.

WACKY FACTS

- The silver and aluminum must be in contact with each other during this experiment because a small electric current flows between them during the reaction.
- Polishing silverware with an abrasive cleanser removes the silver sulfide and some of the silver from the surface. Other chemical tarnish removers dissolve the silver sulfide but also remove some of the silver.
- Baking soda cleanses by neutralizing fatty acids found in most dirt and grease.

The Silver Lining

- The atomic symbol for silver, Ag, is derived from the Latin word *argentums*, meaning "silver." The word *silver* originated from the Anglo-Saxon word *seolfor*.
- Of all the metals, silver conducts electricity best.
- Silver is typically used to fashion jewelry, coins, candlesticks, and silverware.
- Silver is used in dentistry, electronics, and photography.
- Sterling silver is the name for an alloy made of at least 92.5 percent silver by weight. The remaining 7.5 percent is composed of other metals, most typically copper.

TIN

118.710 2 8 18 18 4
Sn
50

44. Bonkers Bonko Cans

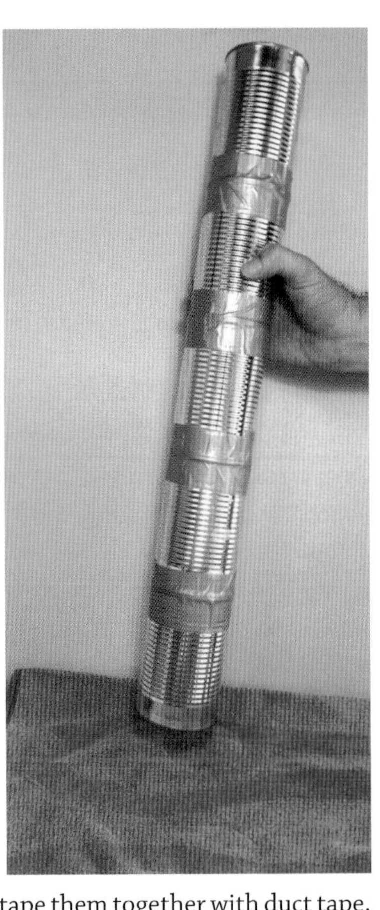

WHAT YOU NEED

- Can opener
- 5 tin cans (all the same size), clean and empty
- Duct tape
- Bath towel

WHAT TO DO

1. Use a can opener to cut off the bottoms of four of the five tin cans.
2. Set the can that retains its bottom on a countertop, with the open end facing upward. Place a second can on top of the first, and tape them together with duct tape.
3. Place the third can on top, and tape it in place.
4. Repeat step 3 with the fourth and fifth cans.
5. Spread the towel on the floor. Holding the can tube with the open end up, bang the bottom of the device against the towel repeatedly, varying the speed and force to produce differing sounds and rhythmic patterns.

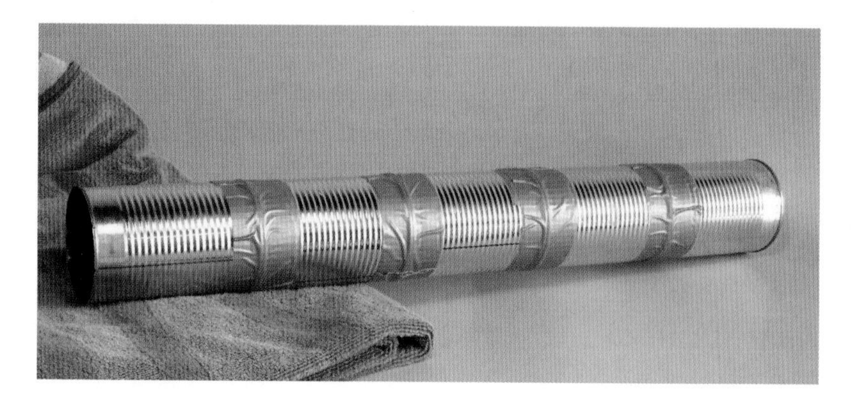

6. While banging the device, cup your hand over the opening to alter the tones.

WHAT HAPPENS

The cans emit strange sounds through the mouth of the tube.

WHY IT WORKS

Vibrations that travel through the air or another medium create sounds. An object vibrating rapidly produces a high-pitched, high-frequency sound. An object vibrating slowly produces a low-pitched, low-frequency sound. When the air inside the tube of tin cans vibrates, the resulting sound contains many different frequencies, which ricochet off the inner walls of the tube, overlapping and combining with one another, amplifying that particular frequency. A long tube of cans amplifies low-frequency sounds. A short tube of cans amplifies high-frequency sounds.

WACKY FACTS

- A bonko made from cans with a wide diameter will sound different from a bonko made from cans with a narrow diameter. Experiment by making a bonko from soup cans and a bonko from tomato or pineapple juice cans.
- Use cans with flat bottoms only. Cans with concave bottoms will not produce the necessary vibrations.
- In Venezuela, people play a musical instrument like a bonko called a *quitiplás*, made from a bamboo tube. Other examples include the *banzara* (which is banged on the ground and hit with sticks) in Tanzania,

ambnuda in Brazil, *dim tenklin* in Malaysia, *juk jang go* in Korea, *ch'ung tu* in China, *ka eke-eke* in Hawaii, *ganbos* in Haiti, and *kendang awi* (two stamping tubes made from bamboo, one larger than the other) in West Java, Indonesia.

Tin: Can It!

- The word *tin* is derived from the Anglo-Saxon language. The atomic symbol Sn comes from the Latin word *stannum*, meaning "tin."
- Tin resists corrosion from water.
- Today, the vast majority of tin is used to make solder, a mixture of tin and lead used to join pipes and to make electronic circuits.
- Most tin cans are actually made from steel with a thin coating of tin.
- The earliest known use of tin occurred around 3500 BCE in the city of Ur in ancient Mesopotamia (modern-day Iraq).
- Pewter, a shiny metallic alloy that looks like silver, consists of at least 90 percent tin, from 2 to 8 percent antimony, and up to 3 percent copper.
- Bronze, an alloy of copper and tin, contains up to 25 percent tin.

126.90447

I

53

2 8 18 18 7

IODINE

45. Wicked Purple Plastic Eater

WHAT YOU NEED

- Newspaper
- Measuring cup
- Cornstarch
- Water
- Glass mixing bowl
- Spoon
- Soup bowl
- Rubber gloves
- Eyedropper
- Tincture of iodine (available at drugstores)
- Ziplock sandwich bag

WHAT TO DO

1. Cover a countertop with newspaper.
2. Mix ½ cup of cornstarch and 2 cups of water in the glass mixing bowl, and stir well.
3. Fill the soup bowl with ½ cup of water.
4. Wearing rubber gloves to avoid staining your skin, use the eyedropper to add drops of iodine to the bowl of water until the water turns golden brown.
5. Pour ½ cup of the cornstarch solution into the ziplock sandwich bag.
6. Seal the bag securely, and place the sealed bag in the bowl with the iodine water so the cornstarch mixture sits beneath the surface.

7. Let sit undisturbed for 30 minutes.
8. Wear the rubber gloves to discard the iodine solution and cornstarch to avoid staining your skin with iodine.

WHAT HAPPENS

After 30 minutes the cornstarch solution turns a purple-blue color.

WHY IT WORKS

The iodine solution diffuses into the plastic bag, coming into contact with the cornstarch solution in the bag. The cornstarch solution turns purple because iodine is an indicator that turns this color in the presence of starch. The plastic bag is selectively permeable, meaning the bag allowed the iodine solution to pass into it but prohibited the starch from leaving it.

This experiment mimics the process of diffusion that occurs through our cells' cell membrane. The molecules in the iodine solution are small enough to pass through the "cell membrane," but the starch molecules are too large to diffuse out of the "cell." Also, the iodine solution diffused into the "cell" from an area of low concentration of solute (in the bowl) to an area of high concentration of solute (inside the plastic bag) in an attempt to balance the solutions—demonstrating the process of osmosis.

WACKY FACTS

- As this experiment demonstrates, a plastic ziplock bag is permeable to iodine, and therefore storing iodine in a plastic bag would result in a leak.
- If the iodine solution is placed in the plastic bag and the starch solution is placed in the bowl, the iodine will diffuse out of the plastic bag and into the bowl, turning the cornstarch a purple-blue color.
- The thyroid, a small, butterfly-shaped gland in the neck, uses iodine to make a hormone that controls metabolism, regulating growth and development. The human body naturally absorbs iodine and stores it in the thyroid gland.
- A cloud of radioactive steam released by a faulty nuclear power plant contains an abundance of iodine-131—a radioactive form of iodine that causes thyroid cancer.
- A shortage of iodine in the human body can cause a goiter, a swelling of the neck resulting from the enlargement of the thyroid gland.

Iodine Makes Everything Fine

- In 1811, French chemist Bernard Courtois discovered iodine in seaweed.
- French chemist Joseph-Louis Gay-Lussac suggested the new element be named iode, from the Greek word for "violet." British chemist Sir Humphrey Davy suggested the name *iodine* to make the name conform to the recently named element chlorine.
- Pure iodine, which occurs as grayish-black crystals, is poisonous if taken internally.
- A wide variety of iodine compounds are used as mild antiseptics.
- When heated, iodine sublimes, changing directly from a solid to a gas without becoming a liquid.

NEODYMIUM

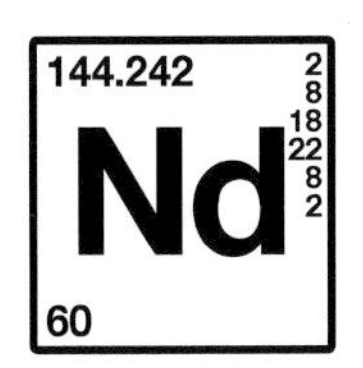

46. Magnetic Cheerios

WHAT YOU NEED

- Soup bowl
- Water
- Cheerios
- 6 Neodymium disk magnets, 0.7 inches diameter by 0.11 inches thick (available at hardware stores)
- Ziplock freezer bag, gallon size
- Measuring cup
- Rolling pin
- Ziplock sandwich bag
- Blow-dryer
- Sheet of white paper

WHAT TO DO

1. Fill the soup bowl with water.
2. Place one Cheerio to float on the surface of the water in the middle of the bowl.
3. Hold the stack of six magnets close to the Cheerio, move the magnets around slowly, and observe the effect on the Cheerio.
4. Fill the ziplock freezer bag with 1 cup of Cheerios.
5. Zip the bag shut most of the way, leaving an opening roughly 1 inch long, and then place your mouth against the opening and inhale to suck the remaining air from the bag.
6. Seal the plastic bag securely, and crush up the cereal into a fine powder by running over the bag repeatedly with the rolling pin.

7. Add 1 cup of warm water to the bag, reseal, and let sit for 15 minutes or until the cereal crumbs turn into goopy mash.
8. Place the stack of magnets inside the empty ziplock sandwich bag, zip the bag shut most of the way, leaving an opening roughly 1 inch long, and then place your mouth against the opening and inhale to suck the remaining air from the bag. Seal the bag securely.
9. Open the ziplock freezer bag, hold the smaller bag inside the larger bag, and gently swirl the smaller bag, moving the protected stack of magnets through the goopy mash for 3 minutes.

10. Remove the small bag from the large bag, and gently dip the bottom of the small bag in the bowl of water to wash any soggy mash residue from the outside of the bag.

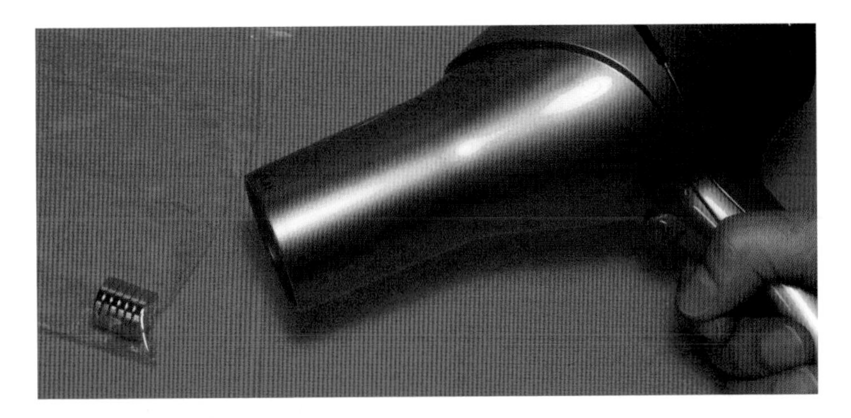

11. Use the blow-dryer to dry the water from the outside of the small plastic bag.
12. Holding the small plastic bag over the sheet of white paper, open the bag, remove the magnets, and gently shake the bottom of the plastic bag over the paper.
13. Inspect the sheet of paper.

WHAT HAPPENS

When you move the magnet around the solitary Cheerio floating in the bowl of water, the Cheerio is drawn toward the stack of magnets. When you examine sheet of paper, you see small black flecks.

WHY IT WORKS

Cheerios contain a significant amount of dust-like iron filings, making the cereal *Os* magnetic. Each cup of Cheerios contains 9.3 milligrams of iron. This small percentage of iron (approximately 2.6 percent) is enough to draw a floating Cheerio toward a stack of strong magnets. Neodymium magnets, made from an alloy of neodymium, iron, and boron, are the most powerful permanent magnets in the world. Crushing the Cheerios and then soaking them in warm water separates out the iron, allowing the stack of strong magnets to pull the specks of iron from the soggy pulp.

WACKY FACTS

- The human body, unable to produce the iron it needs to survive, obtains iron from many iron-rich foods, including meats (beef, pork, turkey), egg yolks, shellfish, produce (beans, raisins, spinach, prunes), and nuts (walnuts, cashews, peanuts).
- Food manufacturers add powdered iron filings (typically elemental or metallic iron) to fortify many cereals (and flour) with iron. While the body cannot absorb elemental iron directly, this powdered iron (and compounds such as ferrous sulfate and ferric phosphate) reacts with hydrochloric acid in the stomach to produces ferrous iron, which the small intestines absorb.
- A healthy adult requires approximately 18 milligrams of iron each day.
- All the iron in the human body (approximately 0.2 ounces) is enough to make two small nails.

Neodymium: The Most Attractive Element

- The name *neodymium* is derived from the Greek words *neos*, meaning "new," and *didymos*, meaning "twin."
- In 1841, Swedish chemist Carl Gustaf Mosander extracted a rose-colored oxide from cerite, which he named didymium, convinced he had discovered a twin of the element lanthanum. In 1885, Austrian chemist Carl Auer von Welsbach separated didymium into two new elemental components: neodymium and praseodymium.
- The ceramic industry uses salts of neodymium to color glass and in glazes.
- Neodymium magnets, the world's most powerful permanent magnet (composed of approximately 70 percent iron, 5 percent boron, and 25 percent neodymium), were simultaneously discovered and created in 1983 by General Motors and Sumitomo Special Metals. The discovery made it possible to miniaturize many electronic devices, including mobile phones, microphones, loudspeakers, and electronic musical instruments.
- Neodymium tarnishes easily, forming a yellow coating when exposed to oxygen in the air.

TUNGSTEN

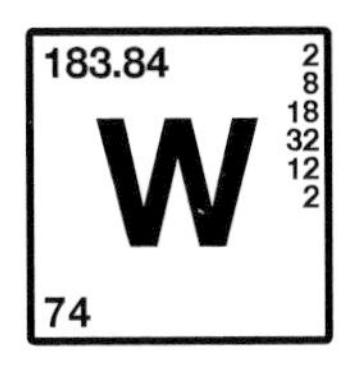

47. Mad Microwaved Lightbulb

SAFETY FIRST

- Safety goggles
- Oven mitt

WHAT YOU NEED

- Microwave oven
- Microwave-safe whisky/juice drinking glass
- Water
- Incandescent lightbulb (with tungsten filament, 60 watts)

WHAT TO DO

1. Wearing safety goggles, remove the rotating tray from the microwave oven.
2. Fill the microwave-safe drinking glass halfway with water.
3. Place the metal end of the incandescent lightbulb into the glass of water. Submerging the metal end of the bulb in water prevents microwaves from causing the metal casing to spark and the lightbulb to explode.
4. Place the glass containing the lightbulb into the center of the microwave oven.
5. Shut the door to the microwave oven, and set the timer for 10 seconds. **Do not set the microwave oven for any longer than 10 seconds**. Otherwise, the lightbulb will heat up too much and explode.
6. Press start, and observe.
7. Wait 2 minutes to allow the lightbulb (which, in a microwave oven, heats up very quickly) to cool down before opening the door, and then wearing an oven mitt and safety goggles, remove the glass from the microwave oven.

WHAT HAPPENS

The lightbulb illuminates.

WHY IT WORKS

The electromagnetic waves pass through the glass of the lightbulb and excite the tungsten filament, pulling the electrons back and forth, causing it to glow.

Note: Heating a lightbulb in a microwave oven irreparably damages the bulb. Do not use the lightbulb in a lighting fixture afterward.

WACKY FACTS

- This experiment works regardless of whether the lightbulb is dead—as long as the tungsten filament remains inside and the glass bulb is intact.
- To test whether your microwave oven is fully sealed or leaking around the door, turn off the lights in the room, place a glass of water in the microwave oven (so you're not heating up an empty oven), turn on the microwave, and hold a fluorescent bulb close to the edges of the door. If the microwave leaks, the tube will light up, meaning you need to get the microwave oven fixed or replaced. Microwaves excite the gas in a fluorescent bulb.
- The lightbulb inside the microwave oven that illuminates the interior is mounted behind a metal grid where very little microwave radiation can reach it.

The Lighter Side of Tungsten

- In 1757, Swedish chemist Baron Axel Frederik Cronstedt coined the term tungsten, a combination of two Swedish words *tung* (meaning "heavy") and *sten* (meaning "stone"), to describe the ore now known as scheelit–discovered in Sweden in 1750.
- Tungsten occurs in nature in the minerals scheelite (calcium tungstate) and wolframite (ferrous-manganous tungstate).
- Two Spanish chemists, Fausto de Elhuyar and his brother Juan José, discovered tungsten in 1783.
- Tungsten has the highest melting point of all metals: 3,410°C (±20°C).
- The atomic symbol for tungsten, W, is derived from the German word *wolfram*, an alternative name for the element. Wolfram is likely a combination of the words *wolf*, referring to the wild animal, and the Middle High German word *rām*, meaning "soot," probably a derogatory allusion to the ore's inferiority to tin.

200.59 2 8 18 32 18 2

Hg

80

MERCURY

48. Totally Tricky Thermometer

WHAT YOU NEED

- Funnel
- Measuring cup
- Plastic 1-liter soda bottle, clean and empty
- Isopropyl alcohol, 90 percent
- Water
- Red food coloring
- Straw
- Modeling dough or modeling clay
- Glass mixing bowl
- Ice cubes

WHAT TO DO

1. Using the funnel, fill ¼ of the plastic soda bottle with 1½ cups of isopropyl alcohol and 1½ cups of water.
2. Add 10 drops of red food coloring into the bottle, and swirl well.
3. Insert the straw into the bottle so that the bottom 1 or 2 inches of the straw are submerged in the liquid and 1 or 2 inches of the straw stick out from the mouth of the bottle.

4. Use the modeling dough or clay to seal the neck of the bottle so the straw stays in place.
5. Place the bottle in the glass mixing bowl, and pour hot water into the bowl until almost full. Observe the red liquid in the straw for 5 minutes.
6. Add ice cubes to the water in the bowl, and observe the red liquid in the straw for 5 minutes.

WHAT HAPPENS

When hot water is placed in the bowl, the red water moves up the straw. When ice is added to the bowl, the red water moves down from the straw.

WHY IT WORKS

Temperature affects the density of liquids. When the alcohol-water solution is warmed by the hot water, the molecules expand and take up more space than they do at room temperature, rising in the straw, just like mercury in a thermometer. Rubbing alcohol boils at a lower temperature than water (180.7°F, as compared to 212°F). When the ice cools down the solution, its volume decreases, going back down the straw.

WACKY FACTS

- Mercury is commonly used in liquid-in-glass thermometers because mercury expands and contracts evenly when heated or cooled, and the element also remains liquid over a wide range of temperatures.
- Alcohol is used in thermometers where the temperature frequently drops below –38°F—the freezing point of mercury.
- A temperature scale is typically printed on the outside of a liquid-in-glass thermometer.

- On the Fahrenheit scale, 32 degrees is the freezing point of water and 212 degrees is the boiling point of water. On the Celsius scale, water freezes at 0 degrees and boils at 100 degrees. On the Kelvin scale, water freezes at 273 K and boils at 373 K.
- Italian astronomer Galileo Galilei invented the first known thermometer in 1593. German physicist Gabriel D. Fahrenheit built the first mercury thermometer in 1714.

Mercury: The Liquid Metal

- The atomic symbol for mercury, Hg, is derived from the silver-colored metal's Greek name, *hydrargyrum*, meaning "liquid silver."
- Mercury is named after the swift messenger of the gods in Roman mythology.
- Mercury is the only metal that is liquid at room temperature.
- No one knows who discovered mercury, but the ancient Chinese, Egyptians, Greeks, Hindus, and Romans knew about the liquid metal.
- In the 19th century, hat makers who used mercuric nitrate to produce felt for hats often succumbed to psychotic symptoms called "mad hatter disease," induced by chronic mercury poisoning. Many people believe that author Lewis Carroll had the condition in mind when he invented the character of the Mad Hatter in his novel *Alice's Adventures in Wonderland.*
- Nearly all fish and shellfish contain traces of mercury, according to the US Environmental Protection Agency. Mercury released into the air through industrial pollution falls from the air and can accumulate in streams and oceans and is turned into methylmercury in the water. Fish absorb the methylmercury as they feed in the water. The mercury builds up more in some types of fish and shellfish than others. The risks from mercury in fish and shellfish depend on the amount of fish and shellfish eaten and the levels of mercury in those fish and shellfish.
- Mercury is a toxic element, and many mercury compounds are extremely poisonous and can cause illness or death. Poisoning can result from mercury vapor inhalation, mercury ingestion, mercury injection, and absorption of mercury through the skin.

LEAD

207.2 2 8 18 32 18 4
Pb
82

49. Ludicrous Lead-Pencil Lightbulb

SAFETY FIRST

- Safety goggles

WHAT YOU NEED

- Wire cutters
- 3 lengths of electrical wire (22 gauge), each 12 inches

- 2 lantern batteries, 6 volt
- 4 alligator clips
- Scissors
- Toilet paper tube
- Electrical tape
- Pie pan
- 0.5 mm mechanical pencil lead refill
- Clear 24-ounce glass jar, clean and empty

WHAT TO DO

1. Using the wire cutters, carefully strip ½ inch of insulation from both ends of all three lengths of electrical wire.
2. Attach one end of one wire to the positive pole on one lantern battery, and attach the free end of that same wire to the negative pole on the second lantern battery.
3. Connect an alligator clip to each of the four ends of the two remaining wires.
4. Using scissors, cut the toilet paper tube in half so the tube measures half its original length.

5. Tape the bottom of one alligator clip attached to one of the wires to one end of the cardboard tube with the clip facing upward. Tape the bottom of one alligator clip attached to the second wire to the same end of the cardboard tube opposite the first clip and facing upward.
6. Stand the cardboard tube in the center of the pie pan and tape it in place.
7. Gently clip the pencil lead between the two alligator clips, suspending it across the top of the cardboard tube opening, being careful not to break the lead and making sure the lead does not touch the cardboard tube.
8. Gently place the glass jar upside down over the toilet paper tube stand.

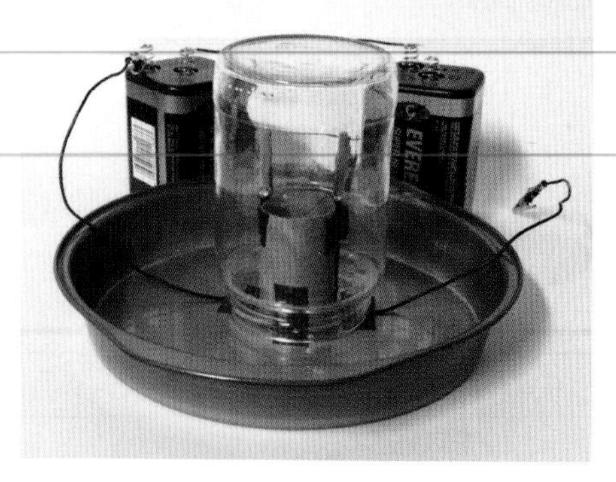

9. Wearing safety goggles in a well-ventilated area, attach one of the free alligator clips to the negative pole on the first lantern battery, and attach the remaining free alligator clip to the positive pole on the second lantern battery.
10. **Do not remove the jar until the glowing filament burns out; otherwise, the lead may catch fire.**

WHAT HAPPENS

The stick of pencil lead illuminates the jar, glowing like a lightbulb.

WHY IT WORKS

The electrons from the battery flow through the pencil lead, which is actually graphite, creating a closed circuit. The graphite or filament becomes so hot that it glows and fills the jar with smoke.

WACKY FACTS

- The lead from a mechanical pencil is actually graphite, which is a form of carbon.
- The cores of wooden pencils have never contained lead. The core is made of nontoxic graphite, a crystalline form of carbon and the highest grade of coal. In ancient Rome, graphite was called *plumbago*, Latin for "a type of lead ore, black lead."
- The jar used to cover the piece of graphite in this experiment prevents oxygen from burning the graphite too quickly.

Get the Lead Out

- The atomic symbol for lead, Pb, is derived from the Latin name for the metal, *plumbum*—the source of the English words *plumber* and *plumbing*.
- The ancient Romans used lead to make plumbing pipes.
- Pure lead is soft and highly malleable. It can be hammered or pressed into thin sheets and permanently stretched without breaking.
- Lead is used as a shield against radiation (due to the metal's high density), for roofing, in stained glass, and to produce lead-acid car batteries.
- Lead from pipes and cookware, leaded gasoline, paints, and fishing weights easily slip into the food chain and get absorbed by the body, resulting in lead poisoning.

BISMUTH

208.98040	2 8 18 32 18 5
Bi	
83	

50. Crazy Crystals

SAFETY FIRST

- Safety goggles
- Respirator mask
- Oven mitts

WHAT YOU NEED

- Aluminum foil
- 2 stainless steel measuring cups
- Metal pie pan
- Hammer
- Bismuth, 1 pound (available from www.rotometals.com)
- Hot plate or stove
- Spoon
- Trivet

WHAT TO DO

1. Crumple up a sheet of aluminum foil into a ball and then form the ball into a cup shape to support and insulate the outside of one of the stainless steel measuring cups.
2. Place the insulated measuring cup in the pie pan.

3. Wearing safety goggles, use the hammer to break the ingot of bismuth into smaller chunks.
4. Place the chunks of bismuth in the second stainless steel measuring cup.

5. Working in a well-ventilated area, place the measuring cup containing the bismuth on the stove and heat to medium high.
6. Place the pie pan containing the insulated measuring cup on a second burner on the hot plate or stove and heat to low.
7. Wait for the bismuth to begin melting, turning up the heat if necessary. The melting point of bismuth is 520.34°F, and this may take 20 minutes or so. **Do not breathe the resulting fumes.**

8. Wearing the oven mitts, gently and carefully shake the measuring cup containing the molten bismuth, looking for ripples in the liquid to verify that the bismuth has melted fully.

9. Use the spoon to carefully pull back the slag (the gray layer floating on the surface of the hot bismuth).
10. Carefully yet quickly, pour the hot liquid bismuth into the insulated measuring cup in the pie pan, leaving the slag in the first cup. Turn off both burners on the stove.
11. Place the measuring cup containing the slag on a trivet in safe spot.
12. Using the spoon, carefully remove the top oxidized layer from the surface of the cooling bismuth in the second cup to reveal the shiny liquid. The top layer affects the color and luster of the resulting crystals.
13. Allow the bismuth to cool down slowly and undisturbed for 5 to 10 minutes. The slower the bismuth cools, the larger the crystals that will form.
14. Still wearing the oven mitts, gently shake the insulated measuring cup to determine if the bismuth has solidified. When you shake the cup, you will see very little rippling in the remaining liquid.
15. Carefully and slowly pour the remaining liquid bismuth from the insulated cup into the first measuring cup—leaving the bismuth crystals behind in the insulated cup.
16. Let the bismuth crystals cool fully.
17. Remove the bismuth crystals from the insulated measuring cup by turning the cup upside down and hitting it gently with the hammer. Crystals will form in the first cup as well.
18. If you're displeased with the bismuth crystal formations, melt the bismuth again and repeat the process.

WHAT HAPPENS

Bismuth crystals form cubic formations with iridescent rainbows of color.

WHY IT WORKS

When you melt the bismuth, the oxides and other contaminants float to the top, forming a layer of slag. When the temperature of the bismuth slowly falls to room temperature, crystals begin forming at the surface, growing downward into the molten metal beneath. Bismuth's solid form floats above its liquid form because liquid bismuth is denser than solid bismuth, the same way ice floats in water. Pouring off the liquid exposes the hot crystals to oxygen, forming a coat of bismuth oxide of various thicknesses on the crystals. The uneven coat of bismuth oxide creates light interference patterns, which produce the iridescent rainbow patterns on the surface of the crystal.

WACKY FACTS

- Bismuth is the active ingredient in Pepto-Bismol. The pharmaceutical industry uses approximately 30 percent of the world's bismuth.
- Bismuth is one of four substances that have a greater density as a liquid than as a solid. The other three are water, gallium, and germanium.
- Bismuth is the most naturally diamagnetic metal, meaning it resists being magnetized and produces a magnetic field directly opposite to an applied magnetic field.

Taking Care of Bismuth

- In 1753, French chemist Claude-François Geoffroy the Younger proved that bismuth, previously considered a form of lead, is a unique element.
- The name *bismuth* is derived from the German words *weisse masse*, meaning "white mass," which were later corrupted to *wisuth* and then Latinized to *bisemutum*.
- Bismuth, found free in nature, is rarer than platinum.

- Because of its low melting point, bismuth alloys are used in fire-alarm systems. Intense heat from a fire turns the bismuth alloy into liquid, triggering the alarms and water sprinklers.
- Alloys of bismuth are used as safety plugs in steam boilers. When the temperature inside the boiler rises too high, these safety plugs melt, allowing the steam to escape before the pressure increases enough to burst the boiler.

IT'S ELEMENTARY
Rockin' with Radium

Polish chemist Marie Curie and her husband, French chemist Pierre Curie, discovered radium in 1898 while experimenting with the mineral pitchblende, a uranium ore. The Curies named the element radium after the Latin word *radius*, meaning "ray"—a reference to the light rays emitted by the silvery metal, causing it to glow in the dark. Radium is an extremely rare element on Earth. Approximately 7 tons of uranium ore are needed to produce a single gram of radium. The radiation given off by radium is more than 1 million times as radioactive as uranium.

Radium was initially used to create glow-in-the-dark paints for use on clocks, watches, and aircraft instruments. Radium was also added to toothpastes, makeup, soap, and candy bars. Eventually, scientists realized that overexposure to radium can destroy tissue in bone marrow that produces red blood cells, causing cancer and eventually death.

In 1934, Nobel Prize-winning physicist Marie Curie died of leukemia, likely caused by the radiation of the radioactive elements (including uranium ore, radium, and polonium) she had been studying during the previous four decades.

Radium slowly decays into radon, then polonium, and finally lead.

Acknowledgments

At Chicago Review Press, I am grateful to my editor, Jerome Pohlen, for sharing my enthusiasm for this book. I am also deeply thankful to copyeditor Julia Loy, designers Andrew Brozyna and Jonathan Hahn, project editor Lindsey Schauer, my agent, Laurie Abkemeier, for suggesting the idea in the first place, researcher and photographer Debbie Green, and my manager, Barb North. Above all, all my love to Debbie, Ashley, and Julia.

Sources

The Basics of Chemistry by Richard Myers (West Port, CT: Greenwood Press, 2003).

Championship Science Fair Projects: 100 Sure-to-Win Experiments by Sudipta Bardhan-Quallen (New York: Sterling, 2007).

Chemical Demonstrations: A Handbook for Teachers of Chemistry, Volumes 1–4 by B. Z. Shakhashiri (Madison: University of Wisconsin Press, 1983–1992).

Chemical Magic by L. A. Ford (Minneapolis, MN: Denison, 1959).

Chemistry Experiments for Children by V. L. Mullin (New York: Sterling, 1961).

"Chicken Eggs" by Rebecca Evanhoe, *Chemical & Engineering News* 84, no. 34 (August 21, 2006): 49.

"Diapers: The Inside Story" by the American Chemical Society, November 18, 2003 .

"Diet Coke and Mentos: What Is Really Behind This Physical Reaction?" by Tonya Shea Coffey, *American Journal of Physics* 76, no. 6 (June 2008): 551–57.

The Dorling Kindersley Science Encyclopedia (New York: Dorling Kindersley, 1994).

The Egotists: Sixteen Surprising Interviews by Oriana Fallaci, translated by Pameal Swinglehurst (Chicago: Henry Regenery, 1963): 239–56.

Einstein's Science Parties by Shar Levine and Allison Grafton (New York: Wiley, 1994).

Encyclopedia of Percussion edited by John Beck (London: Routledge, 1994).

"Evidence for Direct Molecular Oxygen Production in CO_2 Photodissociation" by Zhou Lu, Yih Chung Chang, Qinq-Shu Tin, C. Y. Ng, and William M. Jackson, *Science,* 346, no. 6205 (October 3, 2014): 61–64.

Eyewitness Books: Crystal and Gem by R. F. Symes and R. R. Harding (New York: Knopf, 1991).

"Fluoridation Facts" by the American Dental Association (Chicago: American Dental Association, 2005).

The Handy Chemistry Answer Book by Ian C. Stewart and Justin P. Lomont (Canton, MI: Visible Ink Press, 2014).

Janice VanCleave's 200 Gooey, Slippery, Slimy, Weird & Fun Experiments by Janice VanCleave (New York: Wiley, 1993).

Jr. Boom Academy by B. K. Hixson and M. S. Kralik (Salt Lake City, UT: Wild Goose, 1992).

"Kitchen Experiment Shows the Dangers of Lithium Batteries," *Global News at 5:30 Toronto*, April 2, 2014.

Martin Gardner's Science Tricks by Martin Gardner (New York: Sterling, 1998).

Modern Chemical Magic by John D. Lippy Jr. and Edward L. Palder (Harrisburg, PA: Stackpole, 1959).

Modern Superabsorbent Polymer Technology edited by Fredric L. Buchholz and Andrew T. Graham (Hoboken, NJ: Wiley, 1997).

More Science for You: 112 Illustrated Experiments by Bob Brown (Blue Ridge Summit, PA: TAB Books, 1988).

100 Make-It-Yourself Science Fair Projects by Glen Vecchione (New York: Sterling, 1995).

"Osmosis and the Marvelous Membrane" by Barbara Cocanour and Alease S. Bruce, *Journal of College Science Teaching* 15, no. 2 (1985): 127–30.

"Pharaoh's Serpents" by C. H. Wood, *Chemical News and Journal of Physical Science* 12 (October 13, 1895) edited by Williams Crookes, 170.

Physics for Kids by Robert W. Wood (Blue Ridge Summit, PA: TAB Books, 1990).

Principles of Polymer Chemistry by Paul J. Flory (Ithaca, NY: Cornell University Press, 1953).

"A Protest Against Pharaoh's Serpents," *Journal of the Board of Arts and Manufactures for Upper Canada* 6 (February 1866): 54.

Reader's Digest How Science Works by Judith Hann (Pleasantville, NY: Reader's Digest, 1991).

Science Fair Survival Techniques (Salt Lake City, UT: Wild Goose, 1997).

Science for Fun Experiments by Gary Gibson (Brookfield, CT: Copper Beech Books, 1996).

Science Wizardry for Kids by Margaret Kenda and Phyllis S. Williams (Hauppauge, NY: Barron's, 1992).

Shake, Rattle and Roll by Spencer Christian (New York: Wiley, 1997).

Superabsorbent Polymers by Fredric L. Buchholz and Nicholas A. Peppas (Washington, DC: American Chemical Society, 1994).

Sure-to-Win Science Fair Projects by Joe Rhatigan with Heather Smith (New York: Lark Books, 2001).

365 Simple Science Experiments by E. Richard Churchill, Louis V. Loeschnig, and Muriel Mandell (New York: Black Dog and Leventhal, 1997).

333 Science Tricks & Experiments by Robert J. Brown (Blue Ridge Summit, PA: TAB Books, 1984).

200 Illustrated Science Experiments for Children by Robert J. Brown (Blue Ridge Summit, PA: TAB Books, 1987).

The Way Science Works by Robin Kerrod and Dr. Sharon Ann Holgate (New York: DK, 2002).

What Makes the Grand Canyon Grand? by Spencer Christian (New York: Wiley, 1998).

WEBSITES

"Best How to Make Hot Ice Tutorial (Sodium Acetate)," YouTube video, June 14, 2010, posted by Coolest Science Experiments!, www.youtube.com/watch?v=AedL_NCv1Pw.

"Facts About Hydrogen," Chemicool, www.chemicool.com/elements/hydrogen-facts.html.

"Hooked on Science: 'Black Snake' Experiment" by Jason Lindsey, *Southeast Missourian*, July 3, 2013, www.semissourian.com/story/1983035.html.

"Make Your Own Black Snake Firework," *All Spark Fireworks Blog*, March 28, 2011, www.allsparkfireworks.com/blog/make-your-own-black-snake-firework-noveltites/.

Physics Central, www.physicscentral.com/experiment/physicsathome/all.cfm.

About the Author

Joey Green, the author of the Mad Scientist Handbook series, has demonstrated science experiments on dozens of television shows, including *The View*, *The Other Half*, HGTV's *Smart Solutions*, *Northwest Afternoon*, and *The Debra Duncan Show*.

A former contributing editor to *National Lampoon* and a former advertising copywriter at J. Walter Thompson, Green has written television commercials for Burger King and Walt Disney World, and he won a Clio for a print ad he created for Eastman Kodak before launching his career as a best-selling author. He has written more than 60 published books, including the best-selling Joey Green's Magic Brands series, *Last-Minute Survival Secrets*, *Last-Minute Travel Secrets*, *Contrary to Popular Belief*, and *Clean It! Fix It! Eat It!*—to name just a few.

Green has appeared on dozens of national television shows, including *The Tonight Show with Jay Leno* and *Good Morning America*. He has been profiled in the *New York Times*, *People*, the *Los Angeles Times*, the *Washington Post*, and *USA Today*, and he has been interviewed on hundreds of radio shows.

A native of Miami, Florida, and a graduate of Cornell University (where he founded the campus humor magazine, the *Cornell Lunatic*, still publishing to this very day), he lives in Los Angeles.

VISIT JOEY GREEN ON THE INTERNET AT

WWW.JOEYGREEN.COM